TEXTILE OBJECTIVE MEASUREMENT AND AUTOMATION IN GARMENT MANUFACTURE

ELLIS HORWOOD SERIES IN APPLIED SCIENCE AND INDUSTRIAL TECHNOLOGY

Series Editor: Dr D. H. SHARP, OBE, former General Secretary, Society of Chemical Industry; formerly General Secretary, Institution of Chemical Engineers; and former Technical Director, Confederation of British Industry.

This collection of books is designed to meet the needs of technologists already working in the fields to be covered, and for those new to the industries concerned. The series comprises valuable works of reference for scientists and engineers in many fields, with special usefulness to technologists and entrepreneurs in developing countries.

Students of chemical engineering, industrial and applied chemistry, and related fields, will also find these books of great use, with their emphasis on the practical technology as well as theory. The authors are highly qualified chemical engineers and industrial chemists with extensive experience, who write with the authority gained from their years in industry.

Published and in active publication

PRACTICAL USES OF DIAMONDS
A. BAKON, Research Centre of Geological Technique, Warsaw, and A. SZYMANSKI, Institute of Electronic Materials Technology, Warsaw
NATURAL GLASSES
V. BOUSKA *et al.*, Czechoslovak Society for Mineralogy & Geology, Czechoslovakia
POTTERY SCIENCE: Materials, Processes and Products
A. DINSDALE, lately Director of Research, British Ceramic Research Association
MATCHMAKING: Science, Technology and Manufacture
C. A. FINCH, Managing Director, Pentafin Associates, Chemical, Technical and Media Consultants, Stoke Mandeville, and S. RAMACHANDRAN, Senior Consultant, United Nations Industrial Development Organisation for the Match Industry
THE HOSPITAL LABORATORY: Strategy, Equipment, Management and Economics
T. B. HALES, Arrowe Park Hospital, Merseyside
OFFSHORE PETROLEUM TECHNOLOGY AND DRILLING EQUIPMENT
R. HOSIE, formerly of Robert Gordon's Institute of Technology, Aberdeen
MEASURING COLOUR: Second Edition
R. W. G. HUNT, Visiting Professor, The City University, London
MODERN APPLIED ENERGY CONSERVATION
Editor: K. JACQUES, University of Stirling, Scotland
CHARACTERIZATION OF FOSSIL FUEL LIQUIDS
D. W. JONES, University of Bristol
PAINT AND SURFACE COATINGS: Theory and Practice
Editor: R. LAMBOURNE, Technical Manager, INDCOLLAG (Industrial Colloid Advisory Group), Department of Physical Chemistry, University of Bristol
CROP PROTECTION CHEMICALS
B. G. LEVER, International Research and Development Planning Manager, ICI Agrochemicals
HANDBOOK OF MATERIALS HANDLING
Translated by R. G. T. LINDKVIST, MTG, Translation Editor: R. ROBINSON, Editor, *Materials Handling News*. Technical Editor: G. LUNDESJO, Rolatruc Limited
FERTILIZER TECHNOLOGY
G. C. LOWRISON, Consultant, Bradford
NON-WOVEN BONDED FABRICS
Editor: J. LUNENSCHLOSS, Institute of Textile Technology of the Rhenish-Westphalian Technical University, Aachen, and W. ALBRECHT, Wuppertal
REPROCESSING OF TYRES AND RUBBER WASTES: Recycling from the Rubber Products Industry
V. M. MAKAROV, Head of General Chemical Engineering, Labour Protection, and Nature Conservation Department, Yaroslavl Polytechnic Institute, USSR, and V. F. DROZDOVSKI, Head of the Rubber Reclaiming Laboratory, Research Institute of the Tyre Industry, Moscow, USSR
PROFIT BY QUALITY: The Essentials of Industrial Survival
P. W. MOIR, Consultant, West Sussex
EFFICIENT BEYOND IMAGINING: CIM and its Applications for Today's Industry
P. W. MOIR, Consultant, West Sussex
TRANSIENT SIMULATION METHODS FOR GAS NETWORKS
A. J. OSIADACZ, UMIST, Manchester

Series continued at back of book

TEXTILE OBJECTIVE MEASUREMENT AND AUTOMATION IN GARMENT MANUFACTURE

Editor
GEORGE STYLIOS M.Sc., Ph.D.
Department of Industrial Technology
University of Bradford

ELLIS HORWOOD
NEW YORK LONDON TORONTO SYDNEY TOKYO SINGAPORE

First published in 1991 by
ELLIS HORWOOD LIMITED
Market Cross House, Cooper Street,
Chichester, West Sussex, PO19 1EB, England

A division of
Simon & Schuster International Group
A Paramount Communications Company

Typeset in Times by Ellis Horwood Limited
Printed and bound in Great Britain
by Hartnolls, Bodmin, Cornwall

British Library Cataloguing-in-Publication Data

Textile objective measurement and automation in garment manufacture. —
(Ellis Horwood series in applied science and industrial technology)
I. Stylios, George. II. Series.
677
ISBN 0–13–912742–9

Library of Congress Cataloging-in-Publication Data

Textile objective measurement and automation in garment manufacture /
editor, George Stylios.
p. cm. — (Ellis Horwood series in applied science and industrial technology)
Includes bibliographical references and index.
ISBN 0–13–912742–9
1. Textile fabrics — Mechanical properties — Measurement.
2. Clothing factories — Automation. I. Stylios, George, 1959– . II. Series.
TS1449.T377 1991
687′.04–dc20 91–2212
CIP

Table of contents

Preface 7

PRELIMINARIES

Opening of the conference 13
David J. Johns, Vice-Chancellor and Principal, University of Bradford, UK

Welcome and opening remarks 14
George Stylios, Conference Organizer, University of Bradford, UK

The international dimension 15
Norman Sussman, CBE, President of the British Clothing Industry Association, UK

SESSION 1 The principles of textile objective measurement technology
CHAIR: Professor Ron Postle, Department of Textile Technology, University of New South Wales, Australia

Fabric objective measurement technology: present status and future potential . . . 27
R. Postle, Department of Textile Technology, University of New South Wales, Australia

Objective measurement of the physical characteristics of knitted fabrics 47
D. L. Munden, S. Cassie and L. Ryan, Leicester Polytechnic, UK

The FAST approach to improved fabric performance 55
A. G. DeBoos and D. H. Tester, CSIRO, Division of Wool Technology, Australia

The relation between fibre properties and fabric comfort behaviour 69
K. Slater, School of Engineering, University of Guelph, Ontario, Canada

SESSION 2 Application of textile objective measurement to improve sewing performance of garments
CHAIR: Professor J. Cyril Fumiss, Philadelphia College of Textiles and Science, USA

Recent progress in the application of objective measurement to clothing manufacture 81
S. Kawabata, Department of Polymer Chemistry, Kyoto University,
Masako Niwa, Department of Clothing Science, Nara Women's University,
K. Ito, Osaka Apparel Technology Modernization Association, Osaka, and
M. Nitta,. Oga Clothing Hirakata Co., Osaka, Japan

The use of KES and FAST instruments in predicting processability of fabrics in sewing 107
Herbert Barndt, Fred Fortess, Mel Wiener and J. Cyril Furniss, Philadelphia College of Textiles and Science, USA

Relation between fabric mechanical properties and garment design and tailorability 119
R. L. Shishoo, Swedish Institute of Textile Research, Göteborg, Sweden

An expert system for the prediction of fabric sewability and optimization of sewing and fabric conditions in garment manufacture (sewability system (SS)) . . 139
G. Stylios and J. Fan, Department of Industrial Technology, University of Bradford, UK

SESSION 3 Fabric sewability and garment manufacture
CHAIR: Professor Percy Grosberg, Department of Textile Industries, The University of Leeds, UK

Seam damage predictions and KESF measurements 151
Harriet Meinander, Aija Jylhä and Eeva Kelloniemi, Technical Research Centre of Finland, Tampere, Finland

Relationship between KES properties and sewing performance of difficult-to-sew woven fabrics 163
P. Radhakrishnaiah and Sundaresan Jayaraman, Georgia Institute of Technology, Atlanta, USA

The prediction of garment appearance from measured fabric properties 175
V. Biglia, Vitale Barberis Canonico, Italy,
A. F. Roczniok, CSIRO Division of Wool Technology, Ryde, NSW, Australia
C. Fassima, Gruppo Finanziario Tessile, Italy,
N. G. Ly, CSIRO Division of Wool Technology, Ryde, NSW, Australia

SESSION 4 Robotics and automation in garment manufacture
CHAIR: Professor Paul Taylor, Robotics Research Unit, Department of Electronic Engineering, University of Hull, UK

Robotics applied to the top taping of men's Y-front briefs 191
I. Gibson, P. M. Taylor and A. J. Wilkinson, Robotics Research Unit, Department of Electronic Engineering, Hull University, UK

Sewing dynamics: influence of fabric properties and sewing machine settings on fabric feeding 207
T. J. Little, B. A. Matthews, D. Vass and C. G. Carrere, College of Textiles, North Carolina State University, Raleigh, USA

Stitch quality monitoring in sewing operations 229
J. L. Catchpole and M. Sarhadi, University of Durham, UK

Placing fabric onto moving surfaces 253
M. B. Gunner and P. M. Taylor, Robotics Research Unit, Department of Electronic Engineering, Hull University, UK

SESSION 5 Invited discussion and general conclusions
CHAIR: Dr George Stylios, Department of Industrial Technology, University of Bradford, UK

Index of contributors 279

Subject Index 281

Preface

Perhaps one of the most widely discussed areas of textiles and clothing today is *'fabric objective measurement'*. The term may suggest that it is a textile-related discipline rather than a mainstream clothing one. This may seem true at first sight, the implications of objective measurement, however, are so vast that the subject needs careful consideration, because while fabric is a textile product it is also the main material for clothing. Anything that concerns the behaviour or the characteristics of fabric therefore should be important for both the textile and the clothing industries.

Textile objective measurement is concerned with characterizing textile materials in accordance with their *physical and mechanical properties*. It is the *mechanical properties of fabrics under low stress* which are of primary importance because they are similar to those created under manufacture and wear. This is an important consideration because readywear garments are unlikely to be subjected to *heavy loads*, or to *vigorous bending* and *shear moments*, or to be *compressed under high loads*. The main criterion therefore is that quantifying fabrics by means of their *mechanical properties under low stress* should provide classification of textile structures, which will enable clothing manufacturers to anticipate problems and textile producers to make better products and monitor their process under rigid specifications. It also provides communication between the textile, clothing and allied industries on a scientific basis. This development is in line with *'quick response'*, QR, and *'just in time'*, JIT, principles, which industry has to undertake in order to stay competitive in the marketplace.

Objective measurement has been with us since 1930. The first cited work is that of Peirce who attempted to quantify fabric handle. Lindberg and his co-workers at TEFO in Sweden carried out perhaps the most important work on objective measurement, not only setting the standards, but also being instrumental in developing innovations. Since then various groups at the University of Leeds and the University of Manchester (UMIST) have made more ground in exploring theoretical and experimental methods. Unfortunately these techniques were followed only by academic or research institutes, so their use in the textile and clothing industry had little impact. In the last 20 years Professor Sueo Kawabata in Japan attempted to predict the so-called *'hand value'* of wool-woven suiting material so that fabrics could be engineered with specific properties and assessed for suitability. Kawabata had to devise ways of measuring those important mechanical properties. As a result a line up of instruments was made by Kato Tech, which would enable fabric mechanical properties to be measured under low stress rapidly and accurately. These instruments were applied and tested by Professor Makao Niwa and later commercialized under the name KESF (Kawabata Evaluation System for Fabrics), and include a tensile and shear tester, a pure bending tester, a compression tester and a friction and surface roughness tester. Contemporaneously, and in collaboration with Kawabata, Professor R. Postle and his team in Australia made important breakthroughs in using objective measurement in the *tailorability* of wool-woven suiting fabrics. Recently CSIRO in Australia developed their own version of objective measurement tools named FAST (fabric assurance by simple testing), consisting of a *compression meter, a bending meter*, an *extension meter* and a *dimensional stability test*. Various other groups are now working in this discipline, one of which in the University of Bradford in trying to develop systems which integrate the textile and clothing industries by simplifying objective measurement techniques and using advanced *computer technology, artificial intelligence* and *sensory robotics* to provide *on-line fabric measurements* for the prediction of fabric sewability and optimization of sewing machine conditions for *automated garment manufacture*. The philosophical principle of these systems is to provide interpretation of fabric measurements and implementation of results automatically in the manufacturing process for the industry.

Objective measurement is a very powerful discipline but it is very much dependent upon the correct interpretation and full exploitation of the measured parameters. It must be recognized that objective measurement on its own cannot fulfil all ambitions and solve all problems. It has to be related and used in conjunction with textile and garment manufacture in an interdisciplinary way, in order to deliver its apparent aims. One of these aims is the use of objective measurement in automated garment manufacture and in particular pick and place, feeding and guiding, prediction of sewability, sewing machine automation, etc. If this can be achieved, then *'flexible garment manufacture'*, FGM, becomes a possibility. There is quite a lot of interesting work at present, though it mainly tends to concentrate either in one discipline or in the other. The effects of bridging the two disciplines, textile objective measurement and automation, was the theme of this the first International Clothing Conference with the theme: *'Textile objective measurement and automation in garment manufacture'*.

The conference took place on the 9–11 July 1990 in the University of Bradford and was a truly international event with a cross-section of academics and industrialists from Europe, USA, Asia and Australia, 145 delegates in total, including three distinguished guests: Norman Sussman Esq., CBE, the president of the British Clothing Industry Association who addressed the delegates in a historical speech about the international dimension; Sir Giles Shaw and Mrs Elizabeth J. Peacock, Members of Parliament. The conference consisted of five sessions with seventeen papers in all, as well as an exhibition of prototype equipments from the International

Wool Secretariat, The University of Hull and The University of Bradford. As far as the conference sessions were concerned, we witnessed some stormy ones with constructive discussions throughout the three days. The conference has provided clear indications to the present and future research directions, and I have already undertaken steps to explore some possibilities. This area is of strategic importance for the textile and clothing industries and amongst everything else dissemination of findings and research direction is of paramount importance. To that effect it is my wish to establish this series of international events on a bi-annual basis, and to commit the *International Journal of Clothing Science and Technology* to publish the 'Objective Measurement Technology Bulletin for the Textile and Clothing Industries'.

I would like to thank the European Community for their support, and to express my sincere appreciation to Mr F. Gasperini, to the centre for continuing education of the University, to my editorial board, to Mr R. Deacon and Mr D. Mistry for their invaluable assistance throughout the conference, and to all speakers and delegates from academia and industry who made with their presence and participation this conference, the '1st International Clothing Conference on Textile Objective Measurement and Automation in Garment Manufacture', a great success.

George Stylios
Editor and Conference Organizer

The conference.

Preliminaries

Opening of the conference

Professor David J. Johns
Vice-Chancellor and Principal, University of Bradford, UK

Good morning ladies and gentlemen, may I formally welcome you to this University and wish you all many intellectual exchanges and hope you go away feeling satisfied and that it has been worth your while attending. Looking at the programme and talking to Dr G. Stylios as to the genesis of it, I have no doubt that much good will come out of this meeting.

I do not wish to go through all delegates and mention them by their names, I am sure you will understand why not; but I would like to mention just a few names that have been with us today and to thank them very much for attending this meeting and adding so much to this conference by being here, I have in mind in particular Mr Norman Sussman, CBE, President of the British Clothing Industry Association and also President of L. S. and J. Sussman Ltd; we also have two of our MPs, Mrs Elizabeth J. Peacock and Sir Giles Shaw. I think this is some indication of the importance which the House of Commons and certainly our local MPs give to the clothing industry and the problems associated with it. We thank them very much for finding the time to come this morning. There are other names that I could mention, George has brought attention to the fact that some have travelled from as far away as Japan and Australia. I am referring in particular to Professor Sueo Kawabata from Kyoto University and Professor Ron Postle from the University of New South Wales.

The conference is being supported by the European Community under the BRITE/EURAM programme and for those of you who would like to know a little more about it, please speak with George. In mentioning George's name I would like to commend him for the initiative of getting this programme organized and I also commend him for the initiative of establishing the *International Journal of Clothing Science and Technology* which is now in its second year. I wish the journal every success with the many submitted papers of high quality, which will help all the participants like yourselves to have a forum to submit papers and exchange technical knowledge.

Once again on behalf of the University we welcome you all here most warmly, we hope that you thoroughly enjoy your time here and that you all go away both mentally and physically revived by the events of the day.

Welcome and opening remarks

Dr George Stylios
Conference Organizer, University of Bradford, UK

I would like to welcome you all to the University of Bradford, I am particularly happy to see such an international mixture of delegates from the political life, academia and industry. I would like to mark this as the 1st International Clothing Conference on Textile Objective Measurement and Automation in Garment Manufacture and I very much look forward to our programme.

I am sure that you are as excited as I am to hear our distinguished speaker. I would introduce him by saying that he is the President of the British Clothing Industry Association BCIA, the Chairman of L. S. and J. Sussman Ltd, a major supplier to Marks and Spencer plc for nearer 60 than 50 years, the person responsible for bringing the all seven trade associations into BCIA in 1980, becoming its founding Chairman until May 1987 when he became its first President. He is also the Chairman of the Industry's Wage Negotiating Committee and the Chairman of one of the largest industry pension schemes in the UK, the Industry Pension Scheme, Chairman of the University of Bradford Clothing Option Committee, member of the Confederation of British Industry (CBI), member of the Clothing and Allied Products Industry Training Board (CAPITB), member of the board of the Devon and Cornwall Training and Enterprise Council and Chairman of its Northern Area Board. Ladies and gentlemen it is my honour to welcome Norman Sussman.

The international dimension

Norman Sussman, CBE
President of the British Clothing Industry Association

I am indeed honoured to be invited to address this international assembly, and feel most privileged to enjoy a place on the platform of this distinguished seat of learning with its time-honoured tradition of close links with the textile industry.

The very skyline of this great city reflects an era of British history, for Yorkshire is steeped in the tradition of the textile and clothing industries. It has seen the great days of prosperity, it has suffered the ravages of recession and depression, but now as throughout the long history of the industry, within the county boundaries there are those companies who are taking on and beating the world and will continue to do so.

Textiles and clothing may be less dominant now than they were, nevertheless the industry is and will remain a significant contributor to both the national and local economies.

It is reported that during the 1990s Bradford is to enjoy a £1 billion of investment. That indeed is a reflection of confidence in the future, in which textiles and clothing will continue to play a major part. By your presence at this conference you identify yourself with the confidence which we all share for the continuing significance and importance of the industry worldwide.

The textile and clothing industries in the United Kingdom over the last five years have invested in excess of £2000 million — investment will continue and the prosperity of this region will in the future as in the past reflect the continuing success and prosperity of the industry.

The significance of the timing of this conference should not be lost upon us, for we stand to-day at one of the great turning points in the history of our industry as the forces of international negotiation gather together to seek understanding, to seek conciliation and to seek agreement in the development of world trade.

The potential for international trade in textiles and clothing has never been stronger. Merchandise of high quality and good value is in demand throughout the entire world. The increasing prosperity that we enjoy in Europe, increasing disposable income, an increasing number of men and women at work and the emergence of a better informed and more demanding consumer are benefits that are now being enjoyed on a worldwide basis, and consequently provide substantial opportunities for the European industry.

But those opportunities can be enjoyed across the world, for those opportunities are available not only to the European industry, not only to the industry of the developed world, but those opportunities exist equally for the developing nations as well, provided that consensus can be reached on the framework for international trade following the expiry of the present multi-fibre arrangement in July of next year.

To such a learned audience I do not have to spell out the background to the MFA, many of us have been deeply involved on behalf of the industries we represent since it was first introduced as a derogation of GATT in 1974. it is perhaps worthwhile to remind ourselves of the detail of Article 2, for it is the lack of appreciation of this article which is responsible for so much misguided and ill-informed judgement and comment on the multi-fibre arrangement and its implementation. Article 2 states: 'The basic objective shall be to achieve the expansion of trade, reduction of barriers to such trade and the progressive liberalization of world trade in textile products, whilst at the same time ensuring orderly and equitable development of this trade and avoidance of disruptive effects in individual markets and on individual lines of production in both importing and exporting countries'.

From that article, four objectives are paramount, and they are all central to the environment of international trade before there can be any question of the MFA giving way to a return to normal GATT rules.

(1) the expansion of trade
(2) reduction of barriers
(3) the orderly and equitable development of trade
(4) the avoidance of disruption

After a timespan of sixteen years, the question now has to be addressed: has the MFA succeeded or has it failed? Measured against Article 2 we cannot claim that it has been a success, for disruption has occurred on a considerable scale, the reduction of barriers has been on a limited scale and then only of late, whilst abuse of the MFA over the years has been widespread with counterfeiting, dumping and false origin marking being commonplace.

Yet, on the other hand it would be totally wrong to say that the MFA has failed, for in its overall objective of achieving a fairer distribution of world trade, it has to a large extent succeeded. One only has to recognize the development that has taken place in the Far East to acknowledge the economic prosperity that has been generated by the growth of world trade in textiles and clothing. In particular the least developed countries, the poorest countries of the world who would have lost out to the the dominant suppliers like Hong Kong and Korea have been able to secure a share of world trade in textiles and clothing.

I spoke a few moments ago of the significance of the timing of this conference, for it is this very month that the top-level officials will be meeting in Geneva in order to agree a 'framework' to enable the GATT round to be completed by the end of this year. Geneva July 1990 is crucial to the future of the industry.

The current GATT round of multilateral trade negotiations, otherwise known as the Uruguay round, was launched in Punta del Este in September 1986. At the mid-term review on progress in the round, held in April 1989, it was agreed that the process of integrating trade in textiles and clothing, in other words returning to GATT rules should be 'on the basis of strengthened GATT rules and disciplines'. This means that the future of the MFA was firmly linked to progress in strengthening GATT rules and disciplines on such matters as dumping, subsidies, counterfeiting and tariff and non-tariff barriers.

We are now well into the last decade of this century and it is timely and appropriate to seek to provide the conditions for world trade so as to ensure increasing economic prosperity for all nations as the turn of the century becomes increasingly clearer in our sights.

The whole pattern of world trade is vastly different from what it was when the MFA was introduced in 1974, and provided that the objectives of the Uruguay round can be achieved the phasing out of the MFA can be foreseen.

For the world trading system to embrace GATT rules demands a period of adjustment over many years. Progress will undoubtedly be slow and we will need to be assured every step of the way that there is genuine commitment and honest intent by all the participating nations.

The objectives of the Uruguay round can be stated quite simply: to bring about further liberalization and expansion of world trade but, at the same time, (1) to halt and to reverse protectionism and distortion to trade, and (2) to develop a more open, viable and durable multinational trading system, whilst at the same time outlawing unfair trading practices.

The phasing out of the MFA can only take place as and when strengthened GATT rules and disciplines are firmly and effectively in place. There can be no question of any rundown of the MFA before those conditions have not only been introduced but have been proved by time to be operating effectively.

The importance of linkage is firmly recognized by both the United Kingdom and the EC negotiators and it is worth recalling the words of Mr Tran van Thinh, the Community's permanent representative in Geneva, who stated immediately after agreement was reached on the mid-term review, 'Only the progressive reduction and elimination of the many distortions which are the hallmark of textile regimes to-day will allow the integration process to be gradually achieved'.

That statement clearly reflects the objectives which are shared by the textile and clothing industries of the European Community, for if the trading nations of the world are to enjoy free trade, the principle of fair trade must apply. Free trade must be fair trade.

But to achieve a genuine free trade environment is going to demand a degree of political will and commitment of the highest order. I have never been party to the oft expressed view that the government is indifferent to the fate of the textile and clothing industries and we may be reassured by recent statements by ministers of the Government's firm commitment to the policies which the industry is pursuing. Only time will tell however, how deep and how genuine that political will really is.

The textile and clothing industries, despite their adverse balance of trade, make a substantial contribution to the UK economy — a substantial contribution in terms of employment, in terms of value added and in terms of export achievement. Sufficient grounds to anticipate that the current statements of the Government's firm commitment are maintained throughout the negotiations and that the stated objectives are secured in their entirety.

If a satisfactory settlement is to be reached as in all negotiations there will have to be a degree of give and take by all sides. Textiles is but one of the many areas of negotiation and in order to achieve a successful textile agreement there may have to be concessions in other areas. Flexibility in the Community agricultural policy could well pave the way for the settlement we seek.

The renewals of the MFA which have taken place at each review have had one primary objective — to provide a breathing space to enable the industry to adjust and restructure with a return to normal GATT rules as the ultimate objective.

There can be no doubt whatsoever that in the objective of forcing the industry to restructure the MFA has succeeded. The European clothing industry alone has slimmed down to the extent of 350 000 jobs over the last ten years.

Today the European textile and clothing industries enjoy the most modern production facilities in the world. 22.5% of its spinning machines are open-end rotors, as compared with 20% in the USA, 9% in Japan and 5% in the rest of Asia. Shuttle looms which are the best performing looms at present, account for 70% of its weaving machinery, whereas in the USA and Japan they only account for 50%, and only 38% in the rest of Asia. In the clothing industry, computer-aided design and automated lay-planning and cutting are increasingly becoming commonplace, although assembly operations, thankfully in areas of high unemployment, remain labour-intensive.

Europe 1990 is the world centre of fashion. The great designers of the world are European designers, reflecting the feel, the good taste and the very culture which is inbred through our heritage and through which Europe will continue to lead the world in quality and in value.

The quality and value of the European industry provides the platform upon which its export performance is based and enables it to be the largest exporter in the world. The ability to export to other industrialized markets clearly demonstrates the industry's competitiveness. And yet other markets with the greatest growth potential are effectively closed to Community products. The opening of these markets, the opportunity for all trading nations to share in a fair and equitable manner in world trade is a primary objective of the GATT negotiations.

Much has happened in Europe in recent months. We have been witness to the great drama of the Iron Curtain coming down. Western Europe, flooded by a wave of emotion, hastens to provide support and commitment in the pursuit of economic recovery. That economic recovery will in turn lead to market opportunities of substantial proportions. Those opportunities will be for two-way trade and it is well to remember that the European textile and clothing industries have over recent years made significant sacrifices in supporting the economic growth of the developing countries.

Those sacrifices must not have been made in vain and the market opportunities which may well lie before us will only be capitalized upon if the many distortions to fair trade which now exist are finally and permanently eliminated.

Excessive tariff and non-tariff barriers are one of the great scandals of international trade in textiles and clothing. Tariffs in some of the NICs (newly industrialized countries) are particularly high. 85% in Brazil. In Turkey, whose preferential relationship with the community gives its duty-free access to the European markets, tariffs and other levies are extremely high, reaching 124% in the case of knitted outerwear. Most developed countries, even the USA, apply substantially higher tariffs than the European Community, which of course directly affects our exports. For example exports from this very region are affected by 36% on wool fabric.

Non-tariff barriers such as licensing and administrative obstacles — in some instances administrative sabotage — are an even more insidious barrier to trade than tariffs. The mad cow disease syndrome could have been diagnosed in textiles and clothing over many, many years.

Dumping has plagued the industry throughout the sixteen years of the MFA and the rules and procedures must be tightened — tightened by placing the onus of proof on the supplying country; tightened by closer customs scrutiny of import prices; and tightened by introducing effective measures to control secondary dumping, that is the practice of exporting products made from materials which were themselves bought at dumped prices.

Subsidies are another example of unfair trade with many supplying countries directly assisting the competitiveness of their industries with subsidies for exports, grants for investment and operating aids. GATT rules must be strengthened to prevent subsidies or artificial pricing which distort competition. Failure to control subsidies must be balanced by the introduction of countervailing duties in order to neutralize these distortions.

The growth, the very strong growth over recent years of brand names and of designer labels has led to major problems of counterfeiting and of enforcing trade marks, patents and design copyrights. It is therefore essential that a code, a strong code, on intellectual property be introduced.

The MFA, as I have already acknowledged, has brought great benefits to the developing nations, to the extent that some of those now enjoy newly industrialized country status and have highly developed economies. It is totally unacceptable for them to claim the privilege of developing countries by keeping their markets closed against competition by subsidizing exports and obtaining tariff reductions under the EC's generalized scheme of preference.

If we are to achieve the outlawing of unfair practices, we are entitled to seek substantial improvements in the GATT safeguard clause. Article XIX in its present form is cumbersome and unusable. It cannot be applied against a single source of supply — it has to be applied on a global basis, whilst the compensation factor built into its provisions practically rules out the likelihood of the use of this instrument.

An improved article XIX has to be introduced, it should permit action to be taken in the short term as well as providing general safeguards in the longer term. In the short term the enabling legislation should be in place so as to permit direct action against specific countries without compensation where injury is clearly being suffered. In the longer term there must be satisfactory safeguard provisions. The cumbersome requirement to offer compensation or face retaliation must be reviewed.

None of the changes which we seek will be brought about overnight — the implementation of the decision relating to the opening of markets and the effects of the improvement measures will only be felt gradually and clearly require a transitional system so as to ensure a progressive and gradual evolution which will demand an international surveillance body charged with the responsibility of regularly verifying that all parties have completed with their undertaking.

Liberalization can only come slowly, it can only come when it can be clearly demonstrated that the agreed conditions have been met, and a further move towards free trade can then be considered. The obstacles to free trade remain — tariff reductions — the EC is seeking a substantial reduction of duty rates with a maximum rate of 20% for developed and newly industrialized economies, and a maximum rate of 35% for developing countries. There should be a binding of tariffs, whereby rates cannot be increased over what they are at the time of negotiation. There must be an abolition of all measures inconsistent with GATT.

Whilst I have said that liberalization can come slowly, allow me to be specific and define that period as a minimum of ten years. That period has to be divided into stages and those stages must, I repeat must, be subjected to effective GATT surveillance in order that we may be satisfied that we can move from one stage to the next.

I have already referred to events in Eastern Europe, but even as recently as the mid-term review none of us, but none of us, could have envisaged the great changes that have taken place; already the border controls between East and West Germany have been removed, and last week the deutschmark became the official currency for both Germanies. Our hearts must not be allowed to rule our heads, and any increase in access to our markets from the Eastern European countries must be compensated for by a readjustment to the quota arrangements from other sources.

The United Kingdom clothing industry is totally convinced that the fraudulent labelling of garments is pursued in the Eastern bloc. We have been concerned over many years about the amount of Eastern European products which reach the United Kingdom market and other markets of the Community through West Germany. We have firm reason to believe that as much as 20% of United Kingdom clothing imports from West Germany could originate in East Germany and the rest of the Eastern bloc. There is widespread evidence of unfair pricing of East European exports, either by state subsidies or by artificial exchange rates. These problems have to be addressed, for they create a distortion to the balance of world trade equally as much as those countries who have enjoyed a successful economic take-off, but who still keep their markets closed whilst at the same time provide subsidies for their own industry for production and for exports. For example according to the EC report on Turkey's request for accession to the EEC, their very powerful clothing and textile industries enjoyed massive subsidies amounting to 57% for the developed regions of the country and 104% for the less-developed, top-priority regions.

It all has to stop. But will it? Is protectionism going to be the order of the day, or will common sense prevail; and will there be a meeting of minds in the interest of international trade?

Martin Luther King had a dream; he had a dream because he had a vision. Is this not the moment of history when we too should have a vision. In the spring of this year

President Bush declared that the successful completion of the Uruguay round by December was America's top trade priority. Last month, on her visit to the USSR, the Prime Minister stated, 'the Soviet Union is no longer our enemy'. Can we not have a vision, a vision of world peace and a vision of world trade based on a reformed GATT system.

There is no doubt that the current round of GATT negotiations are the most important that have ever been held, with nearly 100 nations representing more than 85% of world trade working to write a new set of clear and enforceable market opening rules for world trade, and having made a formal commitment to the goal of eventual integration of textiles into GATT.

Fail we cannot, for what has to be achieved is the creation of the trading rules with which we shall enter the 21st century. Fail we cannot, but to succeed common grounds must be established between the EC and the United States of America. It is essential that the closest cooperation should take place between the EC and the USA.

The positions being put forward by the USA, by Australia and most of the developing countries are defensive and are not in the spirit of the Punta del Este declaration. In the United States a global quota system is preferred. This further pressure for trade restriction has resulted in the Textile, Apparel and Footwear Act 1990, covered by two identical bills in the House of Representatives and in the Senate. Those bills propose global quotas which would apply to all countries with the exception of Israel and Canada, with whom the United States has free-trade agreements, and the bills would impose a mere 1% annual growth rate.

The significance of this blatant protectionist approach will not be lost in Europe. There is to be no 'Fortress Europe', there cannot be — there must not be — 'Fortress America'. The American market must be as free and as open to goods from the Community as the market of the community is open to goods from across the Atlantic.

The bills furthermore propose a pilot programme for the auctioning of import licences, the revenue from which would go to the United States Treasury. The idea of a quota auction system in the United States is not new, having originated from a seminar sponsored by the American Association of Exporters and Importers held in May 1984.

However, when the famous 'Jenkins' bill was introduced in Congress in 1985, which aimed at 'significant' cutbacks of quotas for import-sensitive products, the quota auction system was not included, neither was it in the 'Son of Jenkins' bill, although it was included in the amendment to that bill.

If the present bill sees the light of day it will deal a shattering blow to the Uruguay round, and if it were to reach the stage where President Bush has to exercise his veto, let us trust that such an outburst of protectionism would meet the same fate as President Reagan's vetoing of 'Jenkins' and 'Son of Jenkins' achieved. For President Bush not to veto the bill, were it to finally reach him, would be in blatant contradiction of his stated trade priority of achieving a successful completion of the Uruguay round.

Unlike 'Jenkins' and 'Son of Jenkins' the proposed bill does not have the support of the AAMA (the American Apparel Manufacturers' Association), which previously joined forces with ATMI (the American Textile Manufacturers' Institute) in seeking protectionist measures. Just as in the United Kingdom, just as in the member countries of the European Community, so American garment manufacturers have been moving overseas to source some of their production, and quota restrictions would act severely agaisnt their interests. One only has to examine what is available on the American market to recognize the extent of overseas sourcing. With upgraded merchandise being produced in the Far East, it is inevitable that the shift would take place.

This development, growing with increasing impetus over recent years, only serves to endorse my judgement that there are those areas where the MFA has succeeded and fulfilled its objectives, and that there is therefore no reason for any of the developing countries to be seeking a rapid phasing out of the MFA, which over the last sixteen years has provided them with ever-increasing growth at a time when the domestic markets of both Europe and America have grown significantly less than the growth embodied in the quota.

But there can be no question of phasing out the MFA until the necessary strengthening of GATT rules and disciplines has taken place. The phasing out period must be a minimum of ten years. Furthermore, it must be introduced in stages, moving from one stage to the next must be totally dependent on those GATT rules being seen to be working in practice.

Chancellor, ladies and gentlemen, we are clearly at a milestone in the history of our industries. The responsibility for those in whose hands the negotiations lie is indeed of the highest order. There is a great divide between the European and American position. That divide will surely narrow when the significance of strengthening GATT rules and disciplines is fully appreciated, for what we are facing now, what we are addressing, are the opportunities and challenges that will lead us into the next century. It is not only for our generation, but for the generations that follow, that we have to carve out an international trading regime based upon which all nations throughout the free world can trade and prosper.

We gather in Bradford today in the full knowledge of the difficulties and challenges that face our industry; but were we not confident of our ability to overcome those obstacles, this international conference could not have taken place. By our very presence we are committing ourselves and the industry to succeed.

Session 1: The principles of textile objective measurement

CHAIR: Professor Ron Postle
Department of Textile Technology, University of New South Wales, Australia

INTRODUCTION

Good afternoon ladies and gentlemen, I will start this afternoon's session by thanking Dr Stylios for arranging this conference, I know very well the amount of work that goes into the arrangement of conferences and we are all very pleased that George has taken this up himself, to arrange this meeting on the theme of 'textile objective measurement and automation in garment manufacture'.

Few minutes have been set for introduction to this afternoon's session, and I would like this time to try to link the two aspects of the theme: the textile objective measurement on one hand and the automation aspect on the other. You know that there is considerable activity on objective measurement and also considerable activity on automation by the introduction of robotics and various automated aspects of garment manufacture. This conference of course tries to seek common ground between those two areas of work.

If one tries to look at the clothing manufacturing operation it is basically involved with the conversion of initially flat pieces of fabric into a three-dimensional garment, through a whole sequence of operations starting with laying, cutting, sewing, fusing, and pressing. To do this one has to convert a fabric into a three-dimensional smooth unpuckered shape through a very complex sequence of processes. The technology of clothing manufacture is in fact very complex, but one thing is clear and that is to be successful there are many requirements but one of them is that the mechanical and physical properties of fabric to start with have to be consistent to the ultimate product which might be a jacket, a pair of jeans, a suit, a shirt, etc. In recent years, the last decade or more, many researchers have aimed towards defining exactly what those fabric requirements are, how they can be measured and how those measurable properties can be used to help in the various operations of garment manufacture. Of course we've seen very great progress on the automation of the laying and cutting operations and there are a number of research projects in several countries now — Japan, the USA and various European Community countries, including the UK — to automate the handling during the garment manufacturing sequence, to feed these pieces of fabrics into the various machines.

If the ultimate aim of automation is to replace human manipulation of fabrics then we need to try to understand the mechanical interaction between the fabric and the processing machines and to do that we need some form of fabric objective measurement. You all know that fabrics differ greatly in their properties, fabrics of one kind have been manipulated by certain machines, but fabrics of a different kind with different properties, even sometimes fabrics of the same nominal type but of different batch, may have sufficiently different mechanical and physical properties that they cannot be manipulated. Thus these are the sort of problems that we are going to face as we try to automate these operations of clothing manufacture.

Consideration should be given to the whole discipline known as fabric mechanics. Outside the textile and clothing field the discipline of fabric mechanics is very dull and not understood; conventional engineers, for example, do not seem to give fabrics the credit of having their own physical properties; they tend to regard fabrics as limp inert sort of materials like plasticine; you do anything with them, fabrics will not react in any way and do as they are told. Anybody working in the clothing industry knows that this is not true and the discipline of fabric mechanics tries to study, firstly at the fundamental level, the structure and properties of textile materials. Secondly we try to look at the ways in which you can measure the properties, reproducibility, accurately and reliably. That is the first aspect of objective measurement: one needs to have reproducible, reliable systems of measuring fabric properties. Fabrics are not easy to measure scientifically. Garments are easily made but fabric properties are not perfectly uniform, so their own testing systems, procedures and practices must be developed.

Then we try to study the relationships between the properties that we measured and the structure of properties at various levels from a single fibre right through to yarn, fabric and clothing. Then we can look at the fourth aspect the interaction of the relationship between those structure–property relationships and processes in what sort of fibres we use, what sort of yarns we use, the fabric, all of which affect the properties of clothing.

Next we set out to develop an objective measurement system for textile materials; and now to get to this aspect of our conference: a system of measuring scientifically quality and performance of textiles and clothing materials. Traditionally the textile and clothing industries have worked on subjective assessments, assessment made by individual judges, people who have worked all their lives and who have become so experienced with the materials and the clothing aspects that they are able by handling the cloth to assess its quality. Hence they have used words that have become descriptors widely used in the industry.

Well in a modern industry we need to make a whole variety of different textiles and clothing in different designs, we need to respond very quickly to market changes and we have to do this in a way that minimizes or in fact completely removes the chance of any mistakes, which are costly. We therefore need to develop some technologically based system to assess textile and clothing quality and that is what we mean by an objective measurement system; to a scientist the words objective measurement are redundant in the sense that all measurements are objective; we are really contrasting these situations with traditional subjective assessments of assessing the quality and performance of clothing and if we are successful in achieving that then other things can follow. For example, we are able to study the interaction then between fabric properties and the conditions of clothing manufacture; we are able to specify fabrics in objective terms for particular types of clothing and in fact textile manufacturers can then go on to develop new fabrics based on objective mechanical and physical properties because they know the properties that they need for particular clothing and they can construct new fabrics to fulfil this need.

Well one important use of this technology is to assist in the automation of garment manufacture because when clothing manufacture is carried out whether the operations are labour-intensive or automated, we need to know how they interact. This ends a short introduction to the theme of this conference.

Fabric objective measurement technology: present status and future potential

R. Postle
Department of Textile Technology, University of New South Wales, Australia

ABSTRACT

The continuing development of the textile and clothing manufacturing industries depends in no small measure on the successful implementation of reliable objective methods for the *specification, prediction* and *control of fabric quality and performance attributes*. In the last decade, we have seen several notable examples of fabric design and development, and production and quality control in textile processing and clothing manufacture in terms of fabric objective measurement technology. The quality and performance characteristics of fabrics are related to their *low stress mechanical, surface and dimensional properties*. The experimental errors involved in the measurement of these properties are known to be much smaller than the errors in subjective assessment of fabric quality attributes, especially those made by individual judges.

We may define the concept of fabric objective measurement as a necesary and sufficient set of instrumentally measurable parameters which are required to specify the *fabric quality, tailorability and clothing performance.* In this way, fabric objective measurement technology provides a *'fingerprint'* of the fabric quality, tailorability and performance implying that any two fabrics will generally differ at least to some extent in their objectively measurable characteristics. *Fabric objective measurement technology* therefore provides the key for scientific and engineering principles to be used for *fabric specification* and design as well as process control. The most important consequence of the introduction of fabric objective measurement technology will be the promotion of technological communication between various sectors of the textile and clothing industry, research and development workers and all other sectors of industry (e.g. fibre producers, retailing, merchandising, machinery manufacturers) concerned with fibres, textiles and clothing.

INTRODUCTION

The textile and clothing industries have traditionally used subjective assessments of fabric quality by individual judges as the basis of fabric evaluation. It has been shown that [1], although a panel of judges from any one country has a reasonably common notion or concept of what constitutes a desirable fabric handle for textile materials, the ability of any individual judge to assess the handle of a particular fabric is subject to unacceptably large errors. Yet, this is precisely the situation which would apply in most industrial or commercial transactions where individuals must rely on their subjective judgement about the quality or particular fabric samples.

The complex notion of fabric handle may be analysed as the interaction between a number of much simpler fabric quality attributes, e.g. smoothness or sleekness, firmness, fullness, crispness and hardness [1,2]. A list of word descriptors for fabric quality translated from Japanese primary hand expressions is shown in Table 1 and the bipolar fabric quality attributes adopted by the Australian Wool Textile Objective Measurement Executive Committee (AWTOMEC) is given in Table 2.

Table 1 — Fabric word descriptors (translated from Japanese Primary Hand Expressions)

Firm
Sleek
Full
Crisp
Hard

Table 2 — Fabric bipolar attributes (Australian Wool Textile Objective Measurement Committee)

Extensible	—	Inextensible
Springy	—	Limp
Firm	—	Supple
Full	—	Lean
Smooth	—	Rough
Warm	—	Cool

The quality, tailorability and performance characteristics of fabrics are related to their low stress mechanical, surface and dimensionless properties given in Table 3, viz. tensile, shear, bending, compression, surface friction, hygral expansion and relaxation [1,3]. The experimental errors involved in the measurement of most of

Table 3 — Fabric low-stress mechanical, surface and dimensional properties

Tensile
Shear
Bending
Lateral compression
In-plane compression/buckling
Surface friction/roughness
Hygral expansion
Relaxation shrinkage

Characteristic features

- Measurement precision much greater than subjective assessment of fabric quality attributes
- Well established measurement procedures available
- Readily stored in computer database
- Relatively easy to communicate between industry sectors
- Basis for
  - product innovation
  - quality assurance
  - process control

  in textile and clothing manufacture

these properties are known to range from 5% to 12% (for four replicate tests [1]) and are therefore much smaller than the errors involved in subjective assessment of fabric quality attributes, especially those made by individual judges. These fabric mechanical and surface parameters are not difficult to measure, they can be readily stored in a computer database, they are much more precise than subjective opinions about fabrics, they are easy to communicate between industry sectors, and furthermore can be used as the basis for product innovation, fabric specification, quality assurance and process control in both textile and clothing manufacture [4–6].

The principles and techniques for the measurement of fabric mechanical properties are reasonably well established. Data on these properties can provide a quality 'language' that can be readily understood throughout the textile and clothing industries and may ultimately lead to the generation of objective quality standards based on modern scientific and engineering principles.

THE CONCEPT OF FABRIC OBJECTIVE MEASUREMENT

We may define the concept of fabric objective measurement as a necessary and sufficient set of instrumentally measurable parameters which are required to specify the fabic quality, tailorability and clothing performance. In this way, fabric objective measurement technology provides a 'fingerprint' of the fabric quality, tailorability and performance implying that any two fabrics will generally differ at least to some extent in their objectively measurable characteristics.

Fabric objective measurement technology provides the key for scientific and engineering principles to be used for:

(1) optimization of fabric properties to engineer new fabrics of desirable quality and performance attributes for particular end uses;
(2) development of new finishes, finishing agents and fininshing machinery for textile materials;
(3) control of fabric finishing/refinishing to meet fabric mechanical, surface and dimensional property goals;
(4) fabric specification and process control for clothing manufacture;
(5) total fabric development from raw material to tailored garments.

The most important consequence of the introduction of fabric objective measurement technology will be the promotion of technological communication between various sectors of the textile and clothing industries, research and development workers and all other sectors of industry (e.g. fibre producers, retailing, merchandising) concerned with fibres, textiles and clothing [4–6].

CLOTHING MECHANICS

Considerable progress has been made in the study of clothing mechanics, i.e. the relationship between fabric mechanical and physical properties and the cutting, sewing and pressing operations employed during clothing manufacture [3,7]. For example, Fig. 1 shows a schematic representation of the pattern for the shoulder/upper sleeve region of a man's jacket. The seams in this region of the jacket must introduce large double (or three-dimensional) curvatures into the garment which involves a high degree of fabric overfeed in the bias direction (approximately 45 degrees to the warp and weft). Typical levels of fabric overfeed employed during sewing of the seams illustrated in Fig. 1 are given in Table 4 for a man's jacket tailored in Japan and in Table 5 for a similar jacket tailored in Australia. It can be

Table 4 — Typical values of overfeed for a men's suit coat tailored in Japan

Position	Bias angle[a]	Overfeed (%)
Shoulder: front/back (1,2)	Front 74°/ back 70°	8.5
Top sleeve/shoulder		
Front (2,3)	57°/20°	16.5
Back (2,9)	52°/0°	11.0
Sleeve/shoulder (3,4)	20°/4°	4.0

[a] Angle of the line joining points nominated on the pattern with the warp direction of the fabric.

Table 5 — Typical values of overfeed for a men's suit coat tailored in Australia

Position	Bias angle[a]	Overfeed (%)	Strain (%) measured on a tailored coat	
			Overfeed	Underfeed
Shoulder: front/back (1,2)	Front 65°/ back 73°	6.1	2.8	−3.5
Top sleeve/shoulder				
Front (2,3)	46°/27°	6.5	5.2	−1.4
Back (2,9)	54°/5°	11.8	9.5	−2.7
Sleeve/shoulder (3,4)	21°/5°	1.1	1.1	0

[a] Angle of the line joining points nominated on the pattern with the warp direction of the fabric.

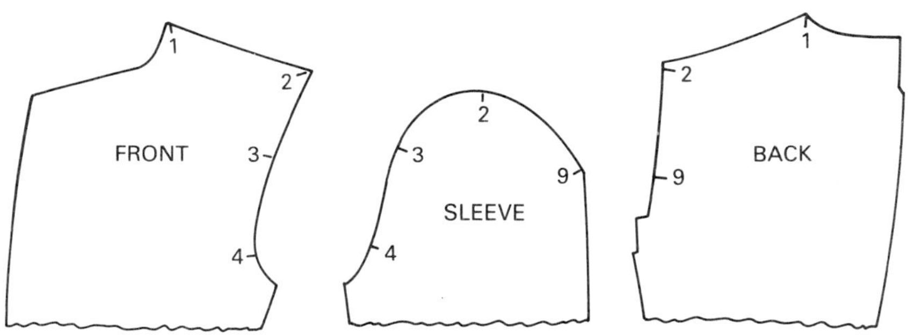

Fig. 1 — Schematic diagram of pattern for a men's jacket showing numbered seams referred to in Tables 1 and 2.

seen from these data that levels of fabric overfeed in excess of 15% are not uncommon in tailoring. Furthermore, it is known that very high localized shear deformations are imposed on the fabric adjacent to these doubly curved seams involving shear angles well in excess of 10 degrees and sometimes in excess of 15 degrees.

It is now generally accepted that fabric tensile, shear, bending and longitudinal compression properties are crucial from the viewpoint of fabric making-up properties or fabric tailorability. There is an optimum combination of these fabric mechanical properties which enable wool fabrics to be satisfactorily tailored into high-quality clothing without any significant puckering, buckling or other unsightly features especially in the seam regions of the garment. This study of clothing mechanics is becoming more important as the level of automation increases during clothing manufacture, a trend which will inevitably proceed during the decade ahead. In order to manipulate successfully fabrics of different mechanical and surface properties, some measurement or sensing device must interact with automated systems during clothing manufacture.

FABRIC MECHANICAL TESTING

The most general method for measuring fabric mechanical properties involves a complete fabric deformation-recovery cycle typical results for which are shown for either tensile or lateral compression properties in Fig. 2 and for fabric shear properties in Fig. 3. In all cases, the deformation-recovery cycle is accompanied by a

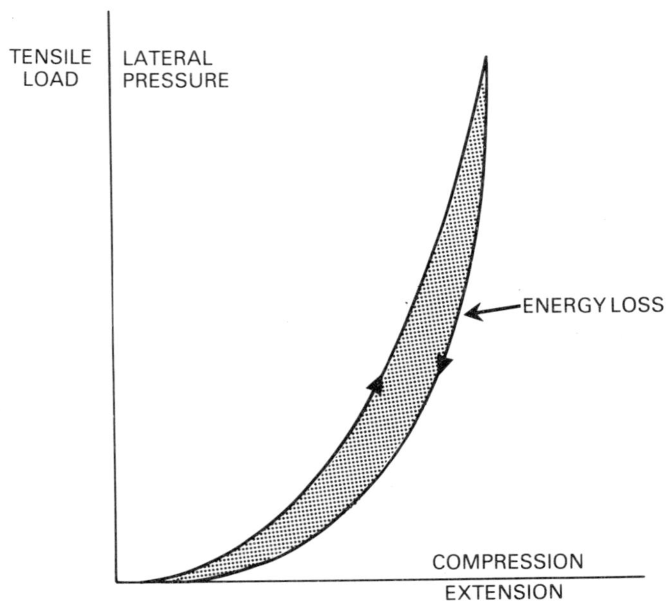

Fig. 2 — Typical deformation–recovery curve for fabric extension or lateral compression, showing the energy loss during a complete deformation–recovery cycle as the shaded area.

significant energy loss or hysteresis as shown in the graphs. From the viewpoint of fabric objective measurement technology, it is possible either the measure the entire deformation–recovery behaviour as is the case for the KESF set of instruments [2] or alternatively, to measure what amounts to a single point on the deformation curve which forms the basis of the FAST set of instruments [8]. The former approach is preferable for research and development work whilst the latter approach has the advantage of simplicity preferable for routine testing purposes.

The shear hysteresis curves shown in Fig. 3 illustrate very clearly the remarkably large influence which fabric finishing has on mechanical properties of wool textile

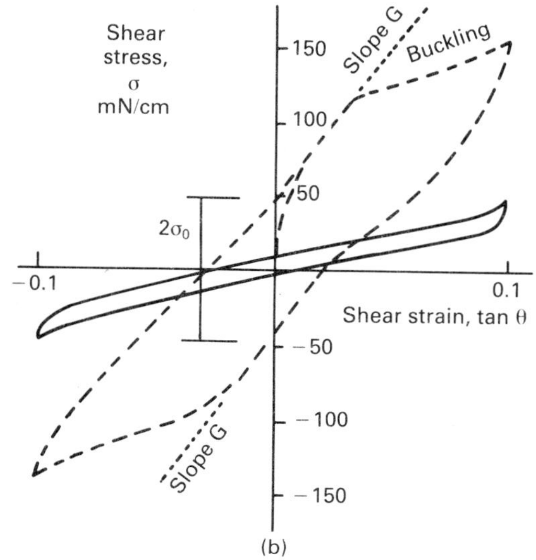

Fig. 3 — Typical shear hysteresis curves of plain-woven fabrics: ----, unfinished; ——, finished.

materials. The crucial part played by fabric finishing in determining these properties is further illustrated in Fig. 4 where the mechanical properties of fabric shear hysteresis, lateral compressibility and surface roughness (weft direction) are shown for one wool fabric and two wool/polyester fabrics after the following stages of finihsing: loom state, heat set (for wool/polyester blends only), scoured, paper pressed and decatized. The potential use of fabric objective measurement technology as a means of controlling wool finishing operations is evident from these results.

Fabric mechanical properties are, of course, influenced not only by fabric finishing but also by the types of wool used for spinning as well as yarn and fabric construction, fabric laundering or dry cleaning and finally by either physical or chemical degradative processes [1]. Measurements of these properties have been successfully used to develop new fabrics based on objectively measured fabric mechanical and physical property goals. The availability of derivative fabric specifications based on the measurement of fabric mechanical, surface and other physical properties provides a clearly defined goal for the development of new fabrics. We shall now look briefly at some specific ways in which fabric objective measurement technology may be utilized for new fabric development, viz. optimization of fabric mechanical and surface properties for suiting materials, some specific examples of the development of new fabric finishes and finally, the concept of total fabric development from fibre raw material through to finished fabric and clothing.

OPTIMIZATION OF FABRIC MECHANICAL AND SURFACE PROPERTIES

Fabric mechanical and surface property measurements can be correlated with subjective assessments of fabric quality attributes such as fabric fullness, sleekness,

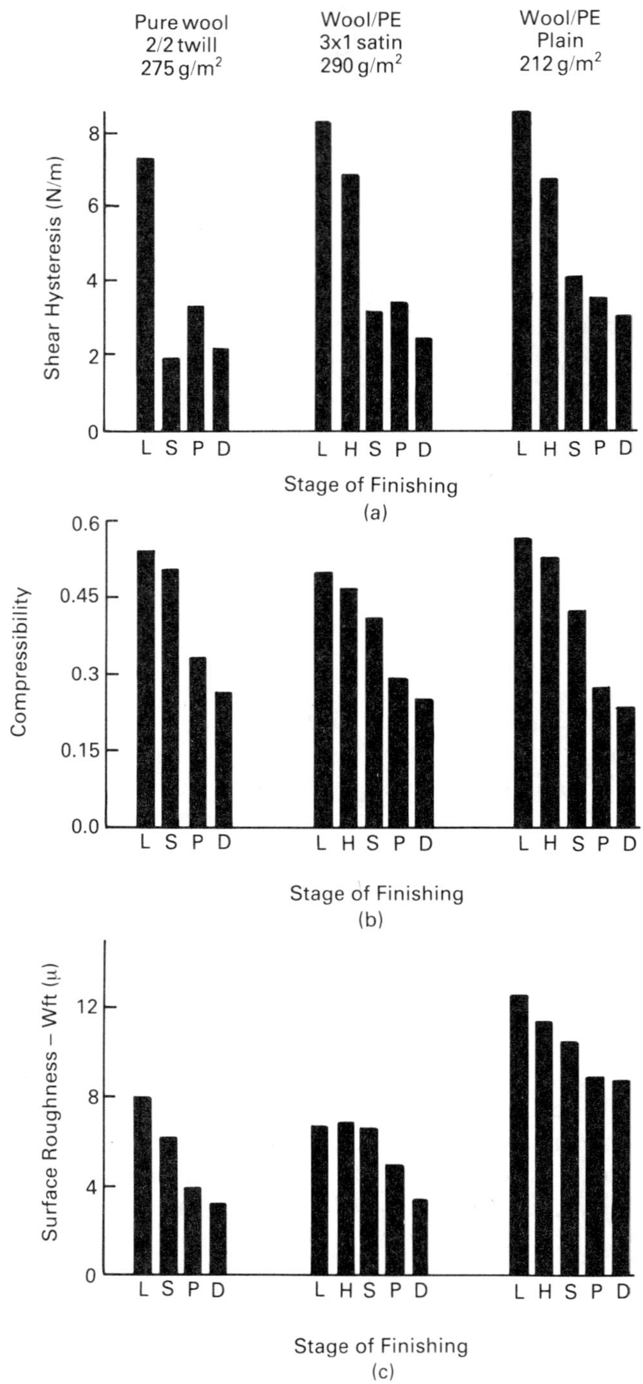

Fig. 4 — Fabric shear hysteresis (a), compressibility (b) and surface roughness in the weft direction (c), after various stages of fabric finishing. L = loom state, H = after heat setting, S = after scouring, P = after pressing, D = after pressure decatizing.

firmess, crispness, hardness and overall fabric handle [1,2]. In Fig. 5, the variation of three of these fabric quality attributes (fabric sleekness, fullness and firmness) as well as the variation of overall fabric handle are shown for the optimum combination of fabric mechanical and surface properties for men's suiting materials ranging in mass per unit area from 220 to 400 g/m^2. These results have been taken from a non-linear

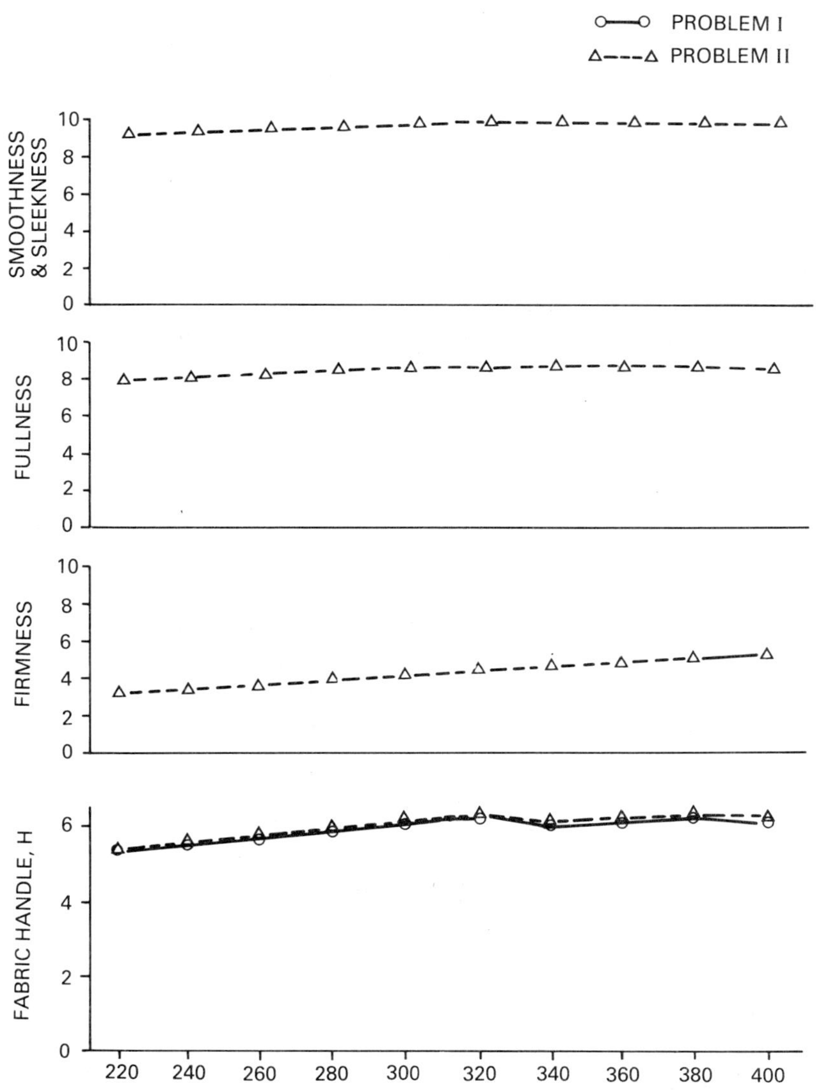

Fig. 5 — The relationship between fabric mass per unit area and fabric quality attributes (or hand values): smoothness and sleekness, fullness, firmness and overall fabric hand for men's winter suiting materials (220–400 g/m^2).

mathematial programnming exercise carried out on an extensive database of fabric measurements for suiting materials [1].

From the results shown in Fig. 5, we conclude that fabrics having the most sought after handle (as assessed subjectively by experts from the textile and clothing industries) are those which have very high ratings of both fabric sleekness and fullness (approximately 9 on a scale of 1–10) and moderate values of fabric firmness (approximately 5–6 also on a scale of 1–10). The combination of fabric mechanical and surface properties required to give these combinations of fabric quality attributes for suiting materials of any particular mass per unit area have also been evaluated.

EXAMPLES OF FABRIC FINISHING

Fabric objective measurement represents a very powerful tool for the control of fabric finishing and refinishing operations, the investigation of additive finishes such as silicones and resin treatments, optimization of decatizing and chemical finishing.

In Fig. 6 a FAST control chart for fabric tailorability is shown. The FAST system of instruments for the measurements of fabric mechanical properties comprise a simple fabric extensometer which is used to extend the fabric in the warp, weft and bias (45° to the warp and weft) directions, a cantilever bending tester and a small fabric thickness/compression tester [8]. The chart shows the results of fabric mechanical property measurements for two wool worsted suiting materials after finishing (the unbroken line labelled A and the broken line labelled B). Difficulties were encountered in tailoring finished fabric A whereas fabric B was found to tailor well. The chart shows on the vertical axis the following fabric mechanical/dimensional property parameters: fabric relaxation shrinkage, fabric hygral expansion (relevant for pure wool fabrics), formability (fabric bending rigidity × fabric extensibility), shear rigidity (calculated from the bias extension of the fabric) fabric thickness, fabric 'surface' thickness (a measure of the compressiblity of the fabric) and finally fabric mass per unit area. Most of these parameters (with the exception of fabric thickness and fabric mass per unit area) have values shown for both the warp and weft directions. It is clearly seen that finished fabric B in Fig. 6 exhibits substantially different measured fabric mechanical properties especially the fabric dimensional and tensile properties leading to a consequent improvement in the tailoring performance of this fabric during clothing manufacture compared with finished fabric A.

Other specific areas related to fabric finishing which have been studied by means of fabric objective measurement technology include the changes in the handle and fabric surface characteristics which can be obtained by the addition of chemicals such as silicones during finishing. Also adverse effects on fabric softness or fabric handle of adding synthetic polymer resins to wool fabrics (for the purpose of shinkproofing) and to cotton fabrics (for the purpose of imparting wrinkle resistant finishes) have also been investigated [4–6].

The finishing operations of steam setting, decatizing and pressing are designed to produce a smooth fabric surface and to remove surface irregularities. Again the

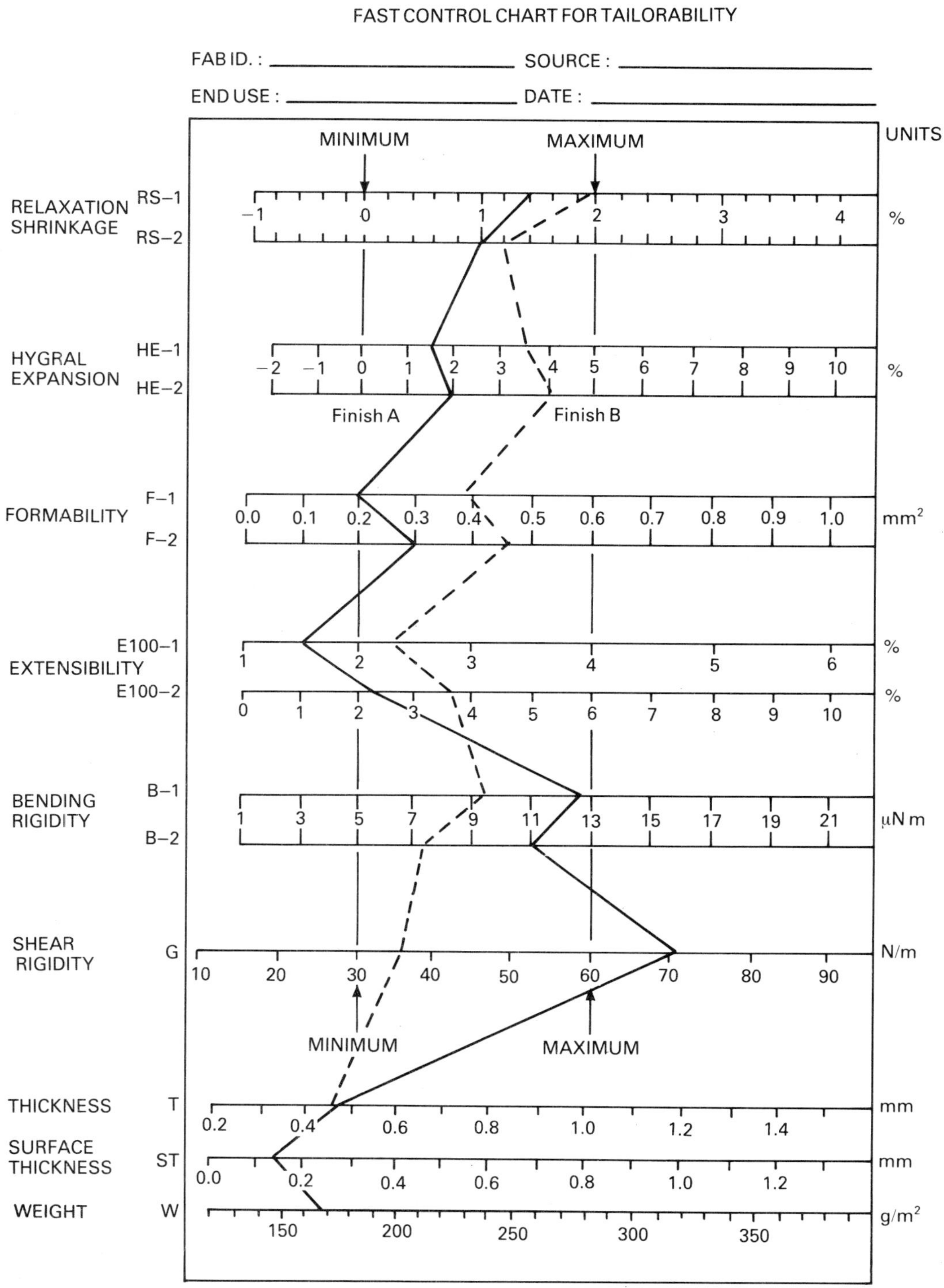

Fig. 6 — Control charts for the fabric tailorability using the FAST instruments (courtesy of Dr A. de Boos, IWS Product Development Centre, UK).

effectiveness of these finishing treatments on the fabric compressibility, surface friction and surface geometry can be measured objectively.

FABRIC DEVELOPMENT

There are a number of instances of new fabric developments based essentially on the measurement of fabric mechanical and surface properties and the consequent attainment of specific objective fabric property goals. Two outstanding examples are the development in Japan of silk-like synthetic materials having extremely soft handle and the development in New Zealand of pure wool worsted crisp tropical light-weight suiting materials [4–6]. We shall discuss here the latter of these two successful fabric developments.

(a) Fabric classification

When considering the possible developments of new fabrics by means of fabric objective measurement technology, it is important to ascertain which definitive class of fabrics is to form the basis of objective comparison for the new fabric being developed. It is important that any such class of fabrics be recognizable in terms of fabric mechanical and physical properties.

A statistical pattern analysis of the measured mechanical and surface properties of a wide variety of men's suiting materials has been reported in the series of Australia–Japan Symposia [6,9]. Table 6 gives a description of two of the fabric classifications which emerge from this statistical analysis; these two fabric classifications are labelled 3 positive and 3 negative — the words 'positive' and 'negative' merely meaning opposites in terms of statistical correlations but not having any connotation of desirable or undesirable features. The three negative fabric classifica-

Table 6 — Factor classification based on staistical pattern analysis of fabric mechanical and surface property data. Only one fabric classification pair (+ 3) is shown

Negative	Positive
	Factor classification 3
Lighter weight	Beautiful, prized fabrics
Clear cut	Cashmere-like
Linen-like, irregular yarn	Casual jacekt or top coat fabric (not suiting)
Harsh, coarse fibre	
'Scratchy'	Thick, bulky
Acceptable summer suitings in twill weave	Poor wear characteristics
Low thickness, low RC, high RT	High RC, low RT

tion for men's suiting materials (shown on the left-hand side of Table 6) is described in terms such as 'light-weight, clear cut, linen-like irregular yarn, low thickness fabrics of low RC (compression resilience) and high RT (tensile resilience)'. The fabric classification labelled 3 positive (on the right-hand side of Table 6) is described as 'beautiful prized fabrics, cashmere-like, casual jacket or top coat fabric, thick bulky fabric of relatively poor wear characteristics, high compression resilience (RC) and low tensile resilience (RT).

The example of a new fabric development discussed below represents a pure wool crisp tropical suiting material which appears to belong to the 3 negative classification.

(b) Pure wool crisp tropical fabrics

In this example of fabric development, the goal is the production of pure wool worsted fabrics having similar properties to crisp wool/mohair tropical suiting materials (about 170 to 180 g/m^2). The project was carried out by the Wool Research Organization of New Zealand in cooperation with the Japanese Hand Evaluation and Standardization Committee [10,11].

The measured mechanical properties of bending, tensile and compression for pure wool worsted yarns are compared in Table 7 with the corresponding properties for wool/mohair blend yarns. The pure wool worsted yarns were produced from a

Table 7 — Yarn properties for New Zealand wool and mohair yarns [10,11]

Yarn	Bending		Tensile				Compression		
	10^3B mN.cm^2	$10^3$2HB mN.cm	LT —	WT mN	RT %	EM %	LC —	10^3WC mN/cm	RC %
A	56.0	20.2	0.62	1.42	76.1	1.88	0.66	113.0	61.7
B	55.7	13.6	0.59	1.00	68.2	1.39	0.57	63.7	56.7
C	46.2	15.5	0.60	1.12	75.6	1.52	0.66	73.6	57.0

A, NZ wool (lot 1)/merino 60/40; B, NZ wool (lot 2)/merino 60/40; C, fine mohair/merino 60/40.

blend of different types of wool — New Zealand crossbred wools (averaging about 35 micrometer diameter) and merino (averaging about 22 micrometres diameter). Thus the strong New Zealand crossbred wool provides similar mechanical properties in the pure wool worsted yarn as the mohair fibres do in the wool/mohair blended yarn.

The fabric quality attributes (or hand values) for men's lightweight tropical suiting produce from the pure wool worsted yarns are compared in Table 8 with the corresponding fabric quality attributes for fabrics of similar weight produced from the wool/mohair blend yarns. In this case, the overall fabric handle and appearance ratings for the two types of fabrics are shown together with the fabric quality attributes (for summer suiting materials) of firmness, crispness, fullness and hardness. It is clear from the results of Table 8 that the fabric quality attributes and the overall fabric handle for the engineered pure wool worsted tropical fabrics are very similar to the corresponding attributes for the model wool/mohair blend fabrics.

Table 8 — Hand values of NZ wool and mohair tropical suitings

Hand value	New Zealand wool				Mohair	
	#7a	#05	#13	#100	Fine #RR1	Kid #RRR1
Handle	3.9	3.8	3.8	3.7	3.4	3.4
Appearance	3.5	3.9	3.7	4.0	4.0	4.4
Primary hand:						
Firmness	4.2	4.5	5.5	4.8	4.6	5.6
Crispness	6.2	5.5	5.9	5.7	5.2	5.7
Fullness	4.5	4.7	5.1	4.5	4.1	4.6
Hardness	4.7	4.9	6.3	5.3	5.6	7.0

Handle = total hand value: 5 — excellent → 1 — poor.
Appearance = total appearance value: 5 — excellent → 1 — poor.
Primary hand values: 10 — very strong → 1 — very weak.

The direct comparison of fabric mechanical and surface properties for the two sets of fabrics is given by the KES data chart shown in Fig. 7. In this data chart, the tensile, bending, shear, compression and surface properties measured on the KES instruments together with the fabric quality attributes (listed in Table 8) are compared directly for the pure wool worsted and the wool/mohair blend tropical fabrics. The goal of producing tropical suiting fabrics by the inclusion of New Zealand crossbred wool in a pure wool worsted yarn has been successful on the basis of comparison with the fabric mechanical and surface properties of wool/mohair yarns and fabrics.

IMPLEMENTATION OF FABRIC OBJECTIVE MEASUREMENT TECHNOLOGY

Several factors need to be considered in respect of the practical implementation of fabric objective measurement technology. These include the organisation of industry based groups, selection of appropriate instrumentation and associated test procedures, database management, development of fabric objective specifications for fabric finishing and appeal manufacture, fabric performance testing and finally potential use of fabric objective measurement in specific textile applications.

(a) Organization of industry based groups

There now exists in Japan, Australia, USA and UK a number of industry groupings which involve academics and research workers with the common aim of evaluating and implementing fabric objective measurement based essentially on the requirements of the local textile, clothing and related industries.

The first group of this kind to be established is the Hand Evaluation sand Standardization Committee (HESC) of the Textile Machinery Society of Japan

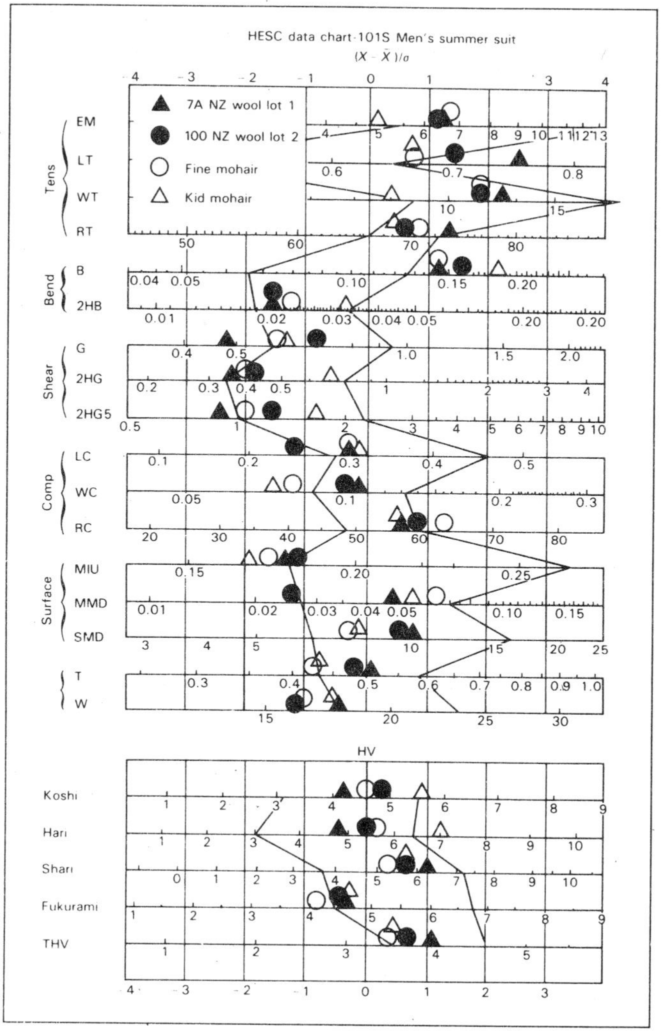

Fig. 7 — Comparison between tropical suitings containing New Zealand wool or mohair (after Kawabata, Carnaby and Niwa [11]). ▲ NZ wool, lot 1. ● NZ wool, lot 2. ○ Fine mohair. △ Kid mohair.

formed originally in 1973. This very strong committee is subdivided into industry groups according to the following classifications: weaving and finishing, men's clothing manufacture, ladies' clothing manufacture, fibre producers and textile machinery manufacturers, and so on. In addition to these companies, membership of the HESC also includes automobile manufacturers, scientific instrument manufacturers, retailing organizations, chemical and soap manufacturers, non-woven fabric manufacturers and so on. This very wide representation of textile, clothing and related companies provides an extremely strong basis for industry communication and implementation of new technology.

The Australian Wool Textile Objective Measurement Executive Committee (AWTOMEC) was formerly convened in 1985 and represents all the major Australian wool interests in Australia including the Australian Wool Corporation, the University of New South Wales, the CSIRO Wool Research Laboratories and the major Australian wool textile and apparel manufacturing companies. The principle aim of this group is to evaluate and implement wool fabric objective measurement technology on an industry-wide basis.

In the USA, two industry groups exist — one based at Philadelphia College of Textile Science and Technology and the other based at North Carolina State University. Both groups comprise the various major research, textile and apparel manufacturing interests in USA. These groups are concerned not only with the implementation of fabric objective measurement in the apparel fabric area but also the development of automated clothing manufacturing systems and also the implementation of fabric measurement in non-apparel areas, e.g. non-woven materials and paper products.

Because of the necessity for all the various sectors of textile and clothing production to liaise actively in the evaluation and implementation of fabric objective measurement technology, it is very desirable that some form of industry based grouping be established in order to achieve success in the dissemination of this technology.

(b) Database management

The successful formation and operation of industry groupings provides a mechanism whereby a database of fabric objective measurements may be established for particular fabric categories of interest to the member companies. For example, the HESC in Japan has already established databases for the objective measurement of the following types of fabrics: men's suiting fabrics, men's jackets, men's slacks, women's suiting fabrics, women's thin dress materials, men's dress shirt fabrics, knitted outerwear fabrics and knitted underwear fabrics. The AWTOMEC group in Australia is steadily building up a substantial database covering wool and wool-rich suiting materials produced from yarns spun on both the worsted and woollen spinning systems.

A data accumulation/analysis system should be linked to instrumentation for the objective measurement of fabrics to produce a fabric data bank. Marketing, processing and subjective evaluation data should also be collated in this fabric data bank. The database should be dynamic in the sense that it is updated regularly by the inclusion of new fabrics, new marketing and processing data and so on. Once enough fabric data has been collated, the data bank should become a very valuable asset for textile and apparel manufacturing and marketing interests.

With the aid of modern computers, the data bank or any specific part of the data may be readily accessed for the purposes of process and production control, quality assurance, fabric design and development and marketing.

The most sophisticated way at present to establish a fabric objective measurement data base would be the use of an expert system which provides a systematic way of storing the knowledge of experts and inferring rules from this knowledge by computerized 'artificial intelligence'. Commercial computer programmes are readily available for the establishment of an expert system which could be used for the implementation of fabric objective measurement technology in the textile and clothing industries.

Other factors involved in the implementation of fabric objective measurement include the selection of instruments, establishment of specifications for clothing manufacture and fabric performance as well as consideration of non-apparel textiles which are covered in detail in the Proceedings of the three Australia–Japan Symposia in this field [4–6].

CONCLUSION

Textile resarch has shown that the mechancial, surface and dimensional properties of fabrics are the most important characteristics which ultimately determine their performance during tailoring as well as the quality of the final garment. The study of the mechanics of textile fibres and structures provides the necessary insight for successful innovation in the fields of fabric testing and metrology, database management, fabric development, process control and quality assurance.

The principles of fibre and fabric mechanics are now being applied to introduce the concept of fabric objective measurement technology whereby the traditional intuitive methods of fabric assessment are replaced by a rational scientific system based on fabric measurements. This new technology is then applied to specific instances of fabric development showing the role of fibre raw material, textile processing and fabric finishing.

We have seen that fabric objective measurement technology can be applied to the development of new fabrics in a number of ways. Firstly fabric properties can be optimized in order to produce fabrics of highly desirable quality attributes.

fabric measurement may be used for the development of new finishes and finishing agents for textile materials such as silicone and synthetic resins. Problems associated with the tailoring of finished fabrics may be overcome by the establishment of fabric mechanical and surface property goals based on fabric objective measurement technology.

As automation proceeds in clothing manufacturing, it will become necessary to adjust fabric handling, cutting, sewing and pressing machinery according to the mechanical and surface properties of the fabric being processed.

Factors which must be considered for the evaluation and implementation of fabric objective measurement technology into the textile and clothing industries include the following: organization of industry-based groups, selection of appropriate instrumentation and associated test procedures, database management, development of fabric objective specifications for textile and apparel manufacture, fabric performance testing and finally potential uses of fabric objective measurement in specific textile applications.

REFERENCES

[1] T. J. Mahar, R. C. Dhingra and R. Postle, *Textile Research J.* Parts I–V: Part I: 1987, **57**, 357–369; Part II: 1989, **59**, 357–368; Part III: 1989, **59**, 448–459; Part IV: 1989, **59**, 721–733; Part V: 1990, **60**, 7–17.

[2] S. Kawabata, *The Standardization and Analysis of Hand Evaluation*, 2nd edn, HESC, Textile Machinery Society Japan, Osaka, 1980.
[3] T. J. Mahar, I. Ajiki, R. C. Dhingra and R. Postle, 'fabric mechanical and physical properties relevant to clothing manufacture', *Int. J. Clothing Sci. and Tech.*, Parts I–III: Part I: 'Fabric overfeed, formability, shear and hygral expansion during tailoring', **1**, 1989, 12–20. Part II: 'Structure balance, breaking elongation and curvature of seams', **2**, 1989, 5–10. Part III: 'Shape formation in tailoring', **3**, 1989, 6–13.
[4] S. Kawabata, R. Postle and M. Niwa (editors), *Objective Specification of Fabric Quality, Mechanical Properties and Performance*, Textile Machinery Society Japan, Osaka, 1982.
[5] R. Postle, S. Kawabata and M. Niwa (editors), *Objective Evaluation of Apparel Fabrics*, Textile Machinery Society of Japan, Osaka, 1983.
[6] S. Kawabata, R. Postle and M. Niwa (editors), *Objective Measurement, Application to Product Design and Process Control*, Textile Machinery Society of Japan, Osaka, 1985.
[7] J. Lindberg, L. Waesterberg and R. Svenson, 'Wool fabrics as garment construction materials', *J. Text. Inst.*, 1960, **51**, T1457.
[8] N. G. Ly, D. H. Tester, P. Buckenham, A. F. Roczniok, M. Brothers and S. de Jong, ITO Report No. 11, Paris Meeting, 1988.
[9] A. E. Stearn, R. L. D'Arcy, T. J. Mahar and R. Postle, 'A classification of fabrics according to their mechanical properties', in Reference [6], 557–566.
[10] G. A. Carnaby, S. Kawabata, M. Niwa, M. Mori, K. Saito and R. J. Walls, 'The utilisation of New Zealand wools in tropical fabrics', in Reference [6], 75–82.
[11] S. Kawabata, G. A. Carnaby and M. Niwa, 'New Zealand/Japan joint project for developing high quality summer suitings using New Zealand wool and fabric objective measurement technology', Proceedings of Advanced Workshop on 'The Application of Mathematics and Physics in the Wool Industry', Christchurch, 1988, pp. 92–103.

QUESTIONS

SARHADI: The properties that you have described, characterize fabric and the resultant quality of clothing. In automated garment manufacture is these set of characteristics sufficient to identify those fabrics or do we have to find other parameters?

POSTLE: I think that in principle it should not make any difference how you are manufacturing a garment whether it is automated or not. In a sense what you are doing is describing a set of properties that are necessary and sufficient, if you like, to convert a flat fabric into let's say a jacket, now whether you do that by hand or whether you do it by machine in an automated way, you still need this combination of properties. The basic difference is in the concept of tailorability. If the fabric is let's say tailored by hand, taking the example of the old bespoke tailor, he can presumably take a lot more care with the fabric, spend much more time on it and eventually tailor the fabric into a satisfactory quality in the end. He can adjust his manipulation with time according to the fabric properties. That is much more difficult of course in an automated system, perhaps possible with advanced sensing devices,but there is a limit to the time. Therefore, I think in principle there should not be any difference, but in practice there will be differences, because an automated tailoring system will have to work rapidly, and what that will mean in the end is that fabrics that are difficult to tailor one way or another may not be able to be produced into a satisfactory quality garment in an automated process. So they may have to either be eliminated or they will have still to be tailored at least partly by hand. Of course this will make them much more expensive.

Introduction to Professor Dennis L. Munden

I would like to introduce the second speaker for this afternoon, Professor Dennis Munden. Dennis Munden started his career in HATRA, and he wrote what I think is one of the classic papers in the textile literature: 'The geometry of plain knitted fabrics', but I think it may today be retitled as, 'Objective measurement of knitted fabrics'. Of course Professor Munden has worked in the University of Leeds and recently retired from Leicester Polytechnic; he is now at the CIMTEX Centre in Leicester. His paper is 'Objective measurement of the physical characteristics of knitted fabrics', co-authored by S. Cassie and L. Ryan.

Objective measurement of the physical characteristics of knitted fabrics

D. L. Munden, S. Cassie and **L. Ryan**
Leicester Polytechnic, UK

ABSTRACT

The commercial development of the *Kawabata Evaluation System (KES)* for the measurement of the *physical properties* of textile materials has led to a concentration on this aspect of textiles to which industry has never previously been accustomed. Knittted fabric applications have not been exempt from this attention.

Previous fundamental studies of the woven structure has revealed the importance of the basic parameters of the structure and their effect on the performance of the material. The Kawabata machinery has been designed to measure the properties and hence correlation between the measurements and the fabric performance is to be expected.

In the case of knitted structures, however, the *critical fabric parameters* are either quite different or, in some aspects, of a different order of magnitude. For these reasons the Kawabata machinery is fundamentally unsuited to determine the *performance of knitted structures*.

The paper will indicate the techniques and measurements necessary in the case of knitted fabrics if information is to be obtained which will correlate with the performance of the knitted structure.

OBJECTIVE MEASUREMENT OF KNITTED FABRICS

No sector of textile material research has received a greater fillip in recent years than the objective measurement of textile materials, as a result of the commercial availability of the KES system developed by Kawabata and Niwa. This has provided textile researchers world-wide with laboratory equipment, designed to measure the critical properties and characteristics of the fabric in a controlled and reliable way, giving results in numerical and graphical form capable of quantitative assesssment and analysis.

The concepts of fabric measurements available from this equipment were based on earlier research work dating from the 1920s to more recent times, including work by Peirce, Lindberg and Grosberg, who had all studied experimentally the theoretical links between the physical properties of the textile materials and their working characteristics. It was only with the development of the KES equipment that reliable standard equipment became available which would permit the properties to be measured in a standard way, so that results obtained in one laboratory might be reproduced and repeated elsewhere.

Much of the early work concentrated on attempts to assess objectively the handle of the textile material, an aspect of considerable economic importance, particularly in the high-value woven suitings made from animal fibres, since this was a characteristic still dependent on the subjective assessment of the skilled buyer, and as such tended to be a matter on which absolute agreement was not usually forthcoming.

Attempts to correlate the measured characteristics of the cloths and their subjective assessments were organized between the wool sectors in Japan and Australia. The significance of this work was not helped by the poor correlation between the alleged subjective measurements from the experts of the two centres and their lack of agreement as to the properties which were subjectively assessed.

More recently, the work has still concentrated largely on woven structures, but the emphasis of much of the work has shifted towards predicting making-up, wearing problems and garment faults and appearance from the 'fingerprint' of the objective measurements of the fabrics.

Thus, pucker, creasing, buckling, tailorability and sewing damage have been shown to be predictable if any of the measured objective properties fall outside defined limits of each material characteristic.

Some limited work has been undertaken on knitted materials, but the results obtained in general have lacked conviction.

A number of reasons can be ascribed to this lack of success. In the first case, the desired features of knitted structures (even if we confine our attention to outerwear fabrics) are very different from those of wovens and many of the properties of the knitted structure are orders of magnitude different from that of woven structures used in similar applications. Air permeability, drape, bending rigidity, extensibility, recovery, hysteresis, fabric thickness, surface irregularity and dimensional stability are some such examples.

Similarly, and because these characteristics are so different, the problems associated with the handling and making-up of the knitted garment are intrinsically different and associated with different fabric characteristics. For example, creasing, seam buckling, pucker, do not concern the manufacture of knitted garments; laddering, loop breaking at the seam, however, do. Flattening of seams, in many aspects of knitted structures is not required. In fact seams are produced with emphasized seam thickness, by use of overlock sewing structures. Whereas in the case of the woven structure, the requirement is to produce a seam of greater extensibility than the woven fabric structure, in virtually all knitted structures the seam is the most rigid part of the garment, so that the garment size in the worn state is determined by the length of the seam, rather than the dimensions of the garment pieces.

In the case of garments from woven structures, rigid fabric needs to retain its shape on the body, so high extensibility and high hysteresis are undesirable since, such properties will cause the garment to rapidly lose its shape. Exactly the opposite is required and expected of knitted structures, so that the garment will conform to the shape of the body on which it is placed and will be comfortable in wear, even though stretched to dimensions different from the natural size of the garment.

Another, and perhaps the most significant difference between the properties of woven and knitted structures, is the characteristic of extension under strain of the fabric in different directions of the fabric.

Many workers in the field of woven structures, have emphasized the significance and importance of the extension of the fabric on the bias and the associated property of fabric shear. This property is of vital significance in the woven structure, since the fabric is very inextensible in the warp or weft direction, having characteristics in these directions not significantly different from plastic materials. The textile-like characteristics of woven structures, the ease with which they bend and drape and their ability to respond to body movement are all associated with the fabric shear characteristic. In addition, since when making-up a garment, sewing thread has to be inserted between the yarns, movement of the yarns comprising the fabric must take place to allow the seam to be introduced without sewing damage. Such yarn movement can only be along the bias. It is not surprising therefore that all researchers in this area have discovered that of all the objective measurements made on the Kawabata equipment, the shear characteristics are the ones which are given the highest weighting to correlate with fabric performance.

In the knitted fabric none of the above argument applies. The fabric is highly extensible in either the coursewise or walewise direction, and sometimes in both, the extension in the bias direction is of no significance, being somewhere between the two values.

In shear, the fabric distorts its loops in a quite different way, from the alteration of the angle of warp and weft in woven structures, and the phenomenon is of no significant importance in wear or fabric processing, since fabric buckling is virtually unknown, and certainly not a common fabric fault; in addition, sewing faults and damage in sewing are associated with other fabric characteristics.

All the above arguments lead to the conclusion that if one was tackling the problem of attempting to assess the performance of knitted fabric and garments, by objective assessment of the fabric, the present Kawabata KES equipment is neither optimally engineered to make the required measurements, not is it designed to measure the preferred properties.

The problem of measuring these objective properties is becoming of commercial significance. Knitted fabrics do suffer damage and distortion during making-up and may be unusable when made-up for certain applications. With the advent of increased automation in the garment manufacturing process, it is becoming vital to be able to detect changes in pattern shapes and sizes to produce a garment of required dimensions, alterations to sewing machine setting and pressing treatments necessary to handle each particular roll of fabric. Advance warning of any such problems is required. Suck-it-and-see techniques will not be appropriate or possible in the factory of the future, nor in the present climate of just-in-time garment engineering.

Analysis of problems associated with handling and making-up of knitted fabric has led to the following proposed list of important fabric characteristics needed to be assessed objectively if the fabric is to be characterized as far as its handling and manufacturing processes, appearance and wear performance are concerned.

(1) Linear extension and recovery coursewise
(2) Linear extension and recovery walewise
(3) Thickness compression and recovery from compression
(4) Air permeability
(5) Yarn/yarn friction
(6) Fabric/fabric friction
(7) Dimensional stability
(8) k_s and k_r values
(9) Stitch length
(10) Weight per unit area
(11) Bending and drape characteristics
(12) Needle penetrating force

Examples given will indicate the progress which is currently being made with development of equipment to measure these characteristics of the knitted structure, so as to give meaningful and reproducible values to establish the modifications to the manufacturing and making-up process necessary to respond to a change in these fabric characteristics.

QUESTIONS

QUESTION: What caused the problem of variation in the mentioned fabrics?

MUNDEN: The problem here was in the brushing treatment. The manufacturer has brushed the fabric under different conditions. Interesting enough that the manufacturer did its own brushing, the garment was not commissioned brushed.

DeARAUJO: It is much better to monitor the knitting operation, in this instance, and not to monitor the finishing operation. It seems that the manufacturer did not control the fabric density.

MUNDEN: I absolutely agree with you. The problem here really is not a garment problem but a textile problem. However, it is the garment manufacturer who has to solve the problem when the fabric is received and it is different in characteristics. It is the garment manufacturer's responsibility whether to go in automation or not, that if differences such as these will appear, they will critically affect the final garment. Hopefully by measuring these properties when the fabric is received, the garment manufacturer could instantly find out which manufacturer produces different than specified knitted fabric.

DeARAUJO: If the width and the weight can be monitored and corrected in a modern stenter then you wouldn't have this difference.

MUNDEN: Absolutely, I could not agree with you more. The answer was that the garment manufacturer did not know that fabric manufacturers were not producing the fabric to the same specifications. This problem came to us because the retailer was aware of the variation that he was getting from different manufacturers. If you are making knitted fabrics then in fact even if all the textile side is put right the making-up operation is still going to be left holding the label. The garment manufacturer needs to facilitate monitoring of these variations.

STYLIOS: As far as garment manufacture is concerned, I wonder how much attention you have paid to the interaction of the sewing operation and the fabric, because during different sewing machine dynamic and kinetic conditions, these aspects interact in different ways, even when you have the same sewing machine type, and what we have found is that at different speeds things do go wrong. Therefore, even if you are statically able to characterize a particular fabric, this may not be necessarily true under a dynamic situation, i.e. during sewing, and I would have thought that it is quite critical to actually look at the whole situation of automation from the point of view of the complexity of structures, whether we call them, woven, knitted or non-woven, and that from the making-up point of view these structures behave differently under static and dynamic conditions. Is your proposed work in agreement with this point of view or is it going to be yet another fabric characterization exercise?

MUNDEN: I am sure that I. Porat is listening to what you are saying with great interest, because this is the area that he will be working, to find out how these characteristics will be different with different sewing conditions. But if you take our example, if one fabric has higher thickness than the other, then what you have to do to get the same dimension is to alter the presser foot pressure. Fundamentally you are right, I am looking at the practical solution.

STYLIOS: As far as yarn to yarn frictional properties are concerned, can't we extract these properties from the frictional component of bending perhaps out of existing testing instruments like the KESF, without having to develop yet another instrument that, in my opinion, will probably serve the same purpose.

MUNDEN: Well, it does mean another modification or a new analysis of the bending results, because in bending you have the frictional characteristics as well as the flexural rigidity, but unfortunately for this sort of fabric the standard bending apparatus on the KESF is a lot to be desired. Knitted fabrics are of different order to woven fabrics and therefore the apparatus designed to measure perfectly bending of woven fabrics, provides results which are very variable and can be altered even by the variation of holding the fabric in the jaws. Surely it may well be a way of doing this by current bending instrumentation, an alternative bending system ought to be developed, however, and what we are now doing is to measure bending properties by other methods and by the KESF and study how reliable the results are by correlating the results of these methods between them. Your idea is very valid and good but we have to find out in practice the crucial properties of knitted fabrics.

SHISHOO: Just a comment on the question that G. Stylios has raised on yarn to yarn friction. I believe that it is well established that if you look at the hysteresis curve in shear or bending, and calculate the shear stress or the bending moment at zero angle, a good measure of yarn to yarn friction can be attained.

MUNDEN: You may well be right but I am very suspicious about any shear measurement on knitted fabrics. Bending and shear are not critical at all in the making up of knitted garments. Those properties that are critically important are length, width and extension. In woven fabrics it is desirable for the garment to remain as it was when it was bought, we are looking at something relatively inextensible and therefore with very low hysteresis. In knitted fabrics high hysteresis is desirable, so that the garment during wear can take up the body configuration. Therefore the very characteristics that are desirable in knitted fabrics are in some cases exactly the opposite to that of tailored suiting. What we must not do in knitted fabrics is to assume that we are looking at the same characteristics.

DeARAUJO: With reference to the bending of knitted fabrics one has to be careful on the preparation of the sample.

MUNDEN: Absolutely, the handling of the fabric has to be zero. One of the difficulties using the KESF is the preparation and placing of the fabric sample in the jaws of the instrument. You get different results because of different handling applied to fabric samples. That has to be standardized.

Introduction to Dr Alan De Boos

Dr De Boos was educated at the University of New South Wales, Australia, where he received his BSc and PhD. He then went on to work at CSIRO in the Division of Wool Technology. For a couple of years he was seconded to IWS in Ilkley, working on textile objective measurement and his paper today is 'The FAST approach to improved fabric performance', co-authored by D. H. Tester.

The FAST approach to improved fabric performance

A. G. De Boos and **D. H. Tester**†
CSIRO, Division of Wool Technology, Australia

ABSTRACT

The FAST system is a set of instruments and test methods developed by the CSIRO in Australia to measure those properties of wool and wool-blend fabrics which affect their *tailoring performance* and the *appearance of tailored garments* in wear.

The FAST system can be used in *fabric development*, to *optimize finishing routes*, to *evaluate new technologies* (spinning systems, finishing machinery), in *buying-control programmes* used by garment makers as well as in *quality control* at all stages of fabric production and use.

Examples are given to illustrate the use of FAST in each of these applications. The FAST system is used to:

(1) assess the effect of a small modification in *fabric structure* on the potential *tailoring performance* of a wool fabric,
(2) evaluate the effect of *modified finishing routines* on those properties which influence the tailoring performance of the fabric, and
(3) compare the properties of fabric processed using the newer developments in *fabric decatizing* with those of fabric processed in conventional pressure decatizing machinery.

Finally, two examples are given where light-weight woven wool fabrics were found to cause problems in tailoring (*poor seam appearance*). Using FAST, the garment maker was able to diagnose the problem in the fabrics, choose between samples with alternative finishes submitted by the supplier, check the bulk against the chosen

† Present address: International Wool Secretariat, Ilkley Development Centre, UK.

sample and *recommend refinishing* when the bulk was found to differ from the agreed sample.

1. INTRODUCTION

In the process of garment making, fabric is subjected to a wide range of stresses and, as a result, is deformed in a number of different ways. These deformations are imposed to form the 3-dimensional shapes that are required in most structured garments. The response of fabric to these strains varies with the type of fabric, its construction and the methods used in finishing. These many variables will determine whether a fabric makes up well into a garment of good appearance or is difficult to tailor and produces garments with relatively poor appearance.

To predict the performance of a fabric in garment making, it is necessary to measure accurately the response of the fabric to the various stresses imposed. While the properties of fabrics that are important in garment-making operations have been known since the 1960s [1], research to improve the understanding of the garment making process has continued up to the present time. Likewise, instrumentation to measure these important properties of fabrics has also been available for many years but was not widely adopted by the garment-making industry in the 1980s [2,3].

Two important developments have led to the wider use of fabric objective measurements by the garment-making industry and, in response, by fabric manufacturers. The first is the continuing trend towards light-weight structured garments and thus light-weight fabrics with all their attendant problems. Garment makers have found that the limitations on fabric properties are more critical with light-weight fabrics and that much stricter tolerances are required. Moreover, in many instances the subjective skills used for the prediction of fabric performance are not adequate for the assessment of light-weight fabrics.

The second development was a coherent set of instruments which measured the important properties of the fabric under the low strain conditions appropriate for the assessment of performance in garment manufacture [2]. These instruments give garment makers the means to measure and assess the potential performance of fabrics objectively and to identify difficult fabrics before they were cut.

Recently an alternative system of instruments has been developed to measure fabric properties [3]. This system was designed to be simple and appropriate for use in an industrial environment. This paper describes these instruments and gives some examples of their use for the improvement of fabric quality.

2. FAST — FABRIC ASSURANCE BY SIMPLE TESTING

The FAST system consists of three simple instruments and a test method.

FAST-1 is a compression meter which measures fabric thickness at two predetermined loads: 0.196 and 9.81 kPa (2 and 100 g/cm^2). From these two measurements, the surface layer thickness of the fabric can be calculated. The surface layer thickness is defined as the difference in fabric thickness at the two loads and is a good index of the effectiveness of milling, raising and pressing operations in finishing. The surface layer thickness is also related to certain aspects of fabric handle, such as smoothness [4]. If the measurements made in FAST-1 are repeated after the fabric has been relaxed in water or steam, the relaxed thickness and surface layer thickness can be determined. Comparison of the initial and relaxed surface layer thicknesses gives a good indication of the stability of the finish on the fabric to the conditions of garment manufacture and wear [5]. Problems such as the reappearance of running marks and bubbling on garment panels can often result from poor finish stability. Such problems can be rectified by refinishing the fabric, but not once the fabric has been cut and the garment is made up.

FAST-2 is a bending meter that uses the cantilever bending principle described in the British Standard method (BS 3356(1961)). The instrument uses photocells to detect the edge of the fabric and the bending length is measured automatically. The bending rigidity of the fabric is calculated from the bending length and the fabric weight (per square metre). Bending rigidity is a useful indicator of the stiffness–suppleness [6] of the fabric and is an important influence on the performance of fabrics in garment-making.

FAST-3 is an extensibility meter which measures the extensibility of the fabric in the warp, weft and bias directions. Warp and weft extensibility are measured under three predetermined loads 4.9, 19.6 and 98.1 N/m (5, 20 and 100 g/cm width). Bias extensibility, which is measured only at a load of 5 g/cm width, is used to calculate the shear rigidity of the fabric. Extensibility and shear rigidity are extremely important determinants of fabric performance in making up. Moreover, shear rigidity is a useful index of the effectiveness of a range of finishing operations, particularly those that permanently set wool fabric [7,8].

FAST-4 is a test method for measuring fabric dimensional stability which determines both components of the dimensional stability of wool fabrics: relaxation shrinkage and hygral expansion. Both components can contribute to problems in garment making and appearance in wear [9]. Poor dimensional stability is widely recognized as a source of poor performance in garment making and many finishers pride themselves on producing fabric that is 'ready-for-the-needle'; that is, having satisfactory dimensional stability.

However, not all applications require stable fabrics, or similar levels of relaxation and shrinkage. For example, the relaxation shrinkage requirements of fabrics for autoclave pleating differ from those of suitings [9]. There are differences in the optimum shrinkage required for light-weight and heavier fabrics. The requirement of engineered dimensional stability demands that finishers be able to measure these properties so that they can adjust their processes accordingly.

As well as those properties of fabrics that are measured directly using FAST, a number of properties may be calculated from the measured values. These properties are shown in Tables 1 and 2 respectively; their relevance to the tailoring operation is described in other publications [10,11].

3. INTERPRETATION OF FAST DATA

While measurement of fabric properties using FAST is a relatively simple process, interpretation of the data in terms of the potential performance of the fabric in

Table 1 — Fabric properties that can be measured using FAST

Measured property	Instrument
Fabric Weight (mass per unit area)	
Compression	
Fabric thickness	FAST-1
Fabric surface thickness	
Relaxed thickness	
Relaxed surface thickness	
Bending	
Bending length	FAST-2
Extension	
Warp extensibility	FAST-3
Weft extensibility	
Bias extensibility	
Dimensional stability	
Relaxation shrinkage	FAST-4
Hygral expansion	

Table 2 — Fabric properties that can be derived from FAST measurements

Derived properties	Calculated from
Bending rigidity	Bending length
	Mass per unit area
Shear rigidity	Bias extensibility
Formability	Bending rigidity
	Extensibility at low loads
Finish stability	Fabric surface thickness
	Relaxed surface thickness

garment manufacture is more difficult. Several approaches to the prediction of tailoring performance of fabrics from their properties have been used. These include:

(a) the use of derived predictors of various degrees of complexity [12,13],

(b) the plotting of fabrics properties on 2- or 3-dimensional axes, on which zones of good and poor performance are defined [1].
(c) the development of fabric fingerprints, the shape of which can be used to predict fabric performance [2].

The FAST system uses a fingerprint as an aid to interpreting the data. This approach is not new; several fingerprints based on the data obtained from the KES-F instruments have been developed as aids to interpretation [14,15]. All define good or 'non-control' zones which indicate when fabric properties are in a range compatible with good tailoring performance.

The FAST control chart, shown in Fig. 1, has grey zones which indicate where potential problems in garment manufacture can be anticipated. These are not accept/reject zones, merely zones which forewarn the garment maker that problems can be anticipated and should be allowed for in garment manufacture. This approach is preferred over alternative techniques because of its simplicity, which enables the gaarment maker to categorize different garment-making problems and to identify, and possibly correct (by re-finishing for example), the property or properties associated with poor fabric performance.

4. REPRODUCIBILITY OF FAST DATA

Effective communication about fabric properties between garment manufacturers and their suppliers requires that the measurements made must be both accurate and reproducible. The instruments must be sufficiently accurate to detect variation within and between fabric pieces. In the FAST system this is achieved by:

(a) ensuring that the instruments are accurate and simple to use, and
(b) using difference measurements at very low strains which compensate for any small distortion in the fabric sample.

An interlaboratory trial [16], involving 13 test laboratories in 6 countries, was conducted to determine the reproducibility of the FAST instruments. A summary of the results of this trial is shown in Table 3.

The trial demonstrated that the errors associated with the measurement compared well with those of other instruments used to predict performance in garment manufacture [17]. The figures quoted are critical differences above which differences are statistically significant.

The original finishing route used on the colour-woven fabric involved a light preset, followed by a scour, crop and continuous autoclave finish. The properties of the final fabric varied with the colour; the darker (chrome-dyed) shades were less extensible than the lighter shades. Because the presence of dye and the pH of the dyeing operation modify the swelling, pH, settability and moisture content of the fibres, dyeing will affect the response of fabrics to finishing processes, which in turn will affect fabric properties.

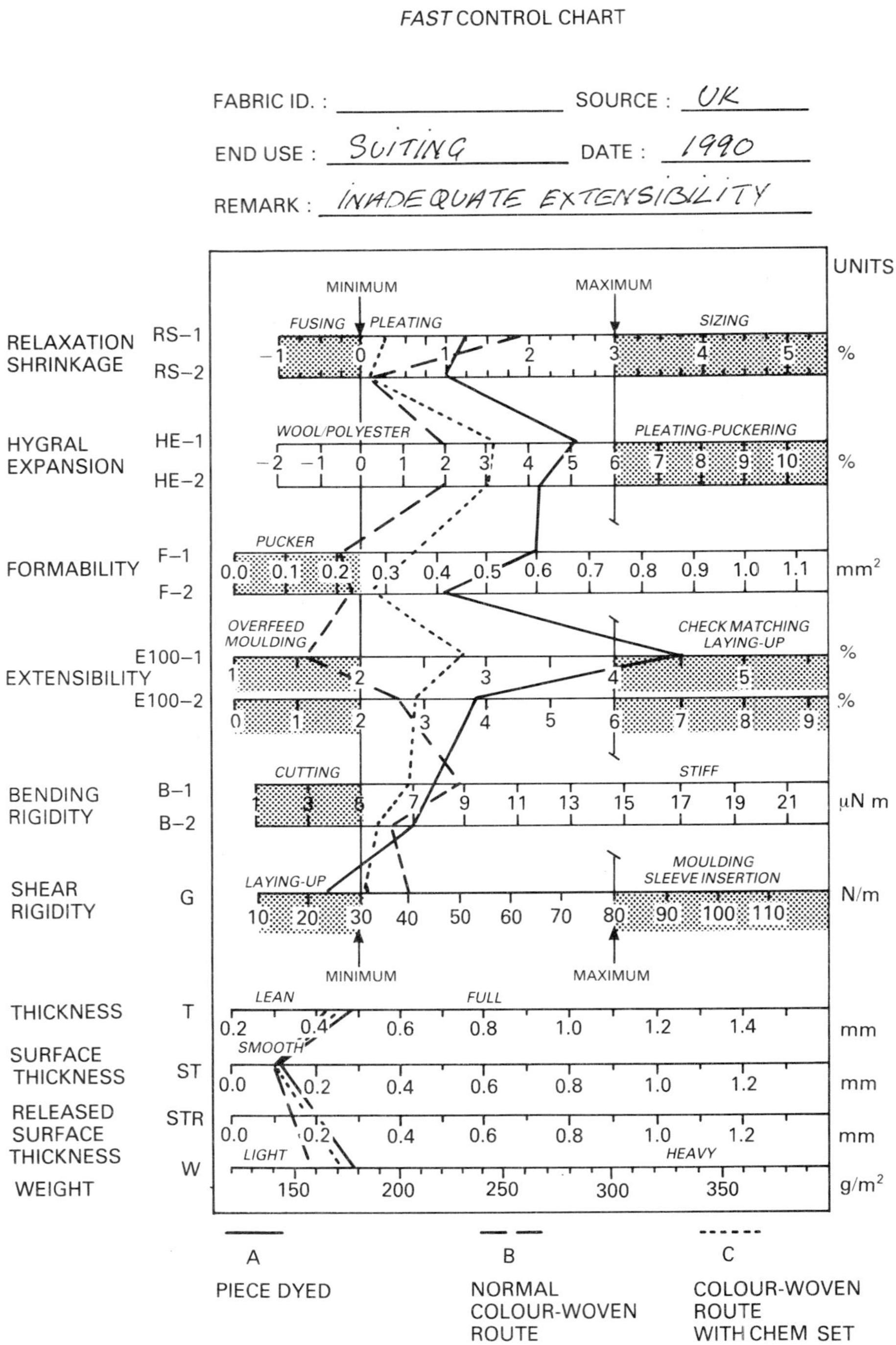

Fig. 1 — Fingerprints of variously finished fabric. A. Piece-dyed. B. Colour-woven (normal conditions). C. Colour-woven with chemical setting.

Table 3 — Summary of results of an interlab trial on FAST

	Critical difference		
	Within lab.	Between lab.	Units
FAST-1			
Thickness at 2 gf/cm^2	0.018	0.031	mm
Thickness at 100 gf/cm^2	0.009	0.026	mm
FAST-2			
Bending length	0.61	1.14	mm
FAST-3			
Extensibility (warp/weft)			
at 5 gf/cm	0.13	0.35	%
at 20 gf/cm	0.15	0.44	%
at 100 gf/cm	0.23	0.70	%
Extensibility (bias)			
at 5 gf/cm	0.26	0.91	%
FAST-4			
Relaxation shrinkage	0.38	0.46	%
Hygral expansion	0.60	0.91	%

To produce a fabric with increased extensibility, it is necessary to increase the crimp in the yarns during finishing and to ensure that the increase is stable to the relaxation processes in garment manufacture and wear. This cannot simply be done by shrinking the fabric in the warp direction (in a stenter or compressive shrinking machine), since the situation the relaxed dimensions of the fabric will be unchanged and negative relaxation shrinkage (expansion) will be introduced into the fabric. This is itself a major cause of problems in garment manufacture.

There are two practical approaches to ensuring adequate fabric extensibility; the first is to control the fabric dimensions in the stenter by limiting the degree to which the fabric is stretched during drying and then to set the fabric in the pressure decatizer. There are difficulties associated with this option. In order to achieve adequate extensibility in the warp, it is often necessary to pull the fabric in the weft, which will reduce weft extensibility. In the fabric shown in Fig. 1, Sample B, the weft extensibility was not high, so that this method could replace one problem with another (less serious) problem.

An alternative approach to achieving adequate extensibility is to wet set the fabric. The mechanism by which wet setting increases yarn crimp in an unset fabric and thus the extensibility (and hygral expansion) of wool fabrics has been explained in a number of publications [18]. By inducing stress relaxation in the fibres, wet setting reduces the inter-yarn forces that exist in a wet, unset (or poorly set) fabric. These forces are responsible for the so-called swelling shrinkage which offsets the effect of hygral expansion when a poorly set fabric is wet out. The dimensional changes that occur when set and unset fabrics are dried are shown in Fig. 2. Inter-

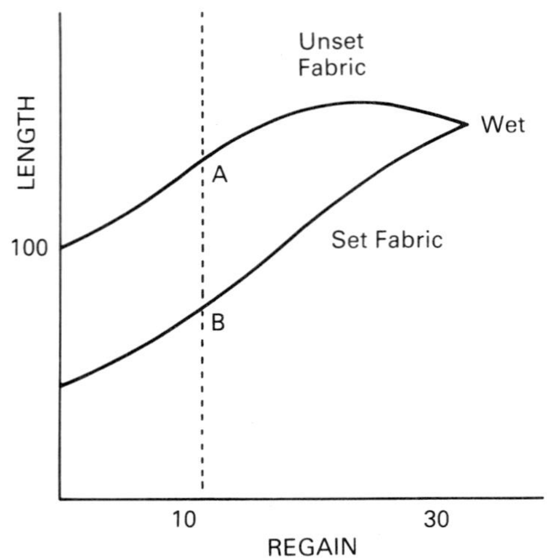

Fig. 2 — Dimensional changes during moisture absorption of unset and wet-set wool fabric.

yarn forces cause the initial expansion observed (between high and intermediate regains) as the unset fabric dries out (Fig. 2). By reducing the inter-yarn forces in the wet fabric, wet setting allows unimpeded shrinkage of the fabric during drying. The apparent shrinkage that results from the wet setting operation is shown in the diagram as line A–B.

The net result is that wet setting causes a permanent reduction in the relaxed dimensions of the fabric, an increase in yarn crimp and an increase in extensibility. The advantage of this approach is that the required relaxation shrinkage can be imparted when the fabric is subsequently dried.

Wet setting can be carried out in the traditional Yorkshire crab, but the tension applied to the fabric can reduce the effect of the setting on the warp extensibility. On the other hand, the permanent set imparted in a continuous crabbing machine (conti-crab) is much lower than that of the Yorkshire crab. Chemical assistants can be used to catalyse setting in the conti-crab.

A chemical setting operation was therefore added to the finishing routine of the colour-woven fabric to engineer the desired properties. The fingerprints are shown in Fig. 1. The extensibility of the colour-woven fabric was considerably higher than that finished by the conventional route. The increased extensibility and formability were reflected in the improved performance of the fabric in garment manufacture. The effectiveness of the alternative setting route depended on the colour of the fabric and thus the pretreatment of the fibres prior to finishing.

As most fabric properties are inextricably linked through the yarn and fabric interactions and configuration, it is inevitable that changing one property will alter others. By increasing yarn crimp and reducing yarn and fibre interactions, wet setting also reduced the bending and shear rigidity of the fabric. However, scope exists for these properties to be subsequently increased by operations that flatten and cohesively set the fabric thereby increasing contact between the yarns at the crossover points. Also the shrinkage in the chemical setting process is reflected in the greater weight per unit area of the resultant fabric.

The modified finishing routine to improve fabric performance in garment manufacture was only possible once the problem had been identified using FAST. Identification of the cause rather than the symptom of poor performance is an essential ingredient of the use of the FAST approach to improved fabric quality. Identification of the cause allows specific targeting of the remedial action.

5. CONCLUSION

In the manufacture of structured garments such as jackets and suits there are values for fabric mechanical properties that will ensure smooth, quick and successful conversion of fabric into finished garments. If the values are outside the optimum ranges for these mechanical properties then a range of options is available to the fabric or garment manufacturer. The garment manufacturer can provide precautionary warnings to the machine operatives or modify the operation where the difficulty will occur. The fabric manufacturrer can transform the properties of the fabric by alteration of finishing processes or refinishing. This paper has demonstrated a combination of the last two alternatives. The final option available is to reject the fabric. Current users of FAST are adopting this only as a last resort.

This demonstration has shown that fabric objective measurement in general, and FAST in particular, has a productive role to play in the textile industry when used in a pro-active way by both fabric and garment manufacturers.

REFERENCES

[1] R. Postle, Proc. 1st Aust/Japan Joint Symp. Obj. Meas., 1982, p. 1.
[2] S. Kawabata, Proc. 1st Aust/Japan Joint Symp. Obj. Meas., 1982, p. 31.
[3] N. G. Ly, D. Tester, P. Buckenham, A. Roczniok, M. Brothers, F. Scaysbrook and S. de Jong, IWTO Conference, 1988, Paris, Report No. 11.
[4] D. Tester and S. de Jong, Proc. Textile Inst. Conf. (Sydney), 1988.
[5] D. Tester, Report to IWTO (in preparation).
[6] F. T. Pierce, *J. Textile Inst.*, 1930, **21**, T377.
[7] E. Finnimore, Proc. 7th Int. Wool Text. Res. Conf. Tokyo, 1985, Vol. 3, p. 80.
[8] Y., Matsui, Proc. 2nd Aust/Japan Joint Symp. Obj. Meas., 1983, p. 301.
[9] T. Shaw, *Amer. Dyestuff Rep.*, 1987, **76**(6), 34.
[10] J. Lindberg, L. Waesterberg and R. Svenson, *J. Textile Inst.*, 1960, **551**, T1457.
[11] N. G. Ly and A. G. De Boos, Proc. 8th Int. Wool Text. Res. Conf. Christchurch, 1990.
[12] M. Niwa, S. Kawabata and K. Ishizuka, Proc. 2nd Aust/Japan Joint Symp. Obj. Meas., 1983, p. 67.

[13] M. Niwa and S. Kawabata, Proc. Adv. Workshop on the Application of Maths. and Phys. in the Wool Ind., New Zealand, 1988, p. 404.
[14] K. Ito and S. Kawabata, Proc. 3rd Aust/Japan Joint Symp. Obj. Meas., 1985, p. 175.
[15] Test Manual of the Australian Wool Textile Objective Measurement Evaluation Committee.
[16] N. G. Ly and A. F. Roczniok, IWTO Conference, 1988, Paris, Report No. 9.
[17] R. C. Dhingra, T. J. Mahar and R. Postle, Proc. 3rd Aust/Japan Symp. Obj. Meas., 1985, p. 803.
[18] K. Baird, *Wool Science Review*, 1867, **31**, 12.

QUESTIONS

QUESTION: Your system does not actually measure hysteresis. Do you consider hysteresis therefore unnecessary for tailorability?

De BOOS: In designing the system we recognized the fact that hysteresis was not measured. In measuring hysteresis we require more complex instrumentation. Part of the aim of FAST is to produce simple instrumentation, appropriate for industry. However, in our experience we have found very few fabrics, none at all in the UK, where the tailoring problem was associated with excessive or insufficient hysteresis in the fabric. We did not design the instruments to measure hysteresis but we have found that this does not cause a problem. There is an argument that says shear hysteresis is very important, we have recognized that in some instances people have associated shear hysteresis to become a problem; however, in the type of fabrics we have been measuring there is a very good correlation between the shear rigidity and shear hysteresis. Obviously we are picking up those fabrics in our measurements of bias extensibility which is related to shear rigidity measurement.

QUESTION: Is bias extensibility a simple translation from shear hysteresis?

De BOOS: From bias extensibility we measure shear rigidity, we found a very good correlation in most wool fabrics between shear rigidity and shear hysteresis.

STYLIOS: We found that the mechanism of seam pucker is based upon the interaction between the sewing thread and the fabric under sewing dynamic conditions. And from the industry's point of view the most difficult group of fabrics prone to seam pucker are man-made light-weight fabrics of the 'papery' type. I wonder whether there is any consideration as far as the interaction between the sewing thread, the fabric, the machine dynamics and the needle with the particular measurements under FAST or is it really that we are talking about instrumentation for property, characterization of wool-type materials only?

De BOOS: Obviously in sewing a seam there is an interaction between sewing thread, the needle and the fabric which is related to seam pucker. In FAST we are measuring the fabric's contribution to seam pucker; the ability of the fabric to absorb in plane forces that lead to seam pucker and that is the only part of the interaction that we actually measure, i.e. the fabric's contribution to the problem. However, we found that formability which predicts seam pucker very well for wool fabrics, also can be used for non-wool fabrics as well. Obviously our experience with CSIRO is with wool but we found formability of non-wool fabrics also to be a very good indication of seam pucker. The fabric is only one part of the equation, but that is the part we try to measuure.

ASSUSA: We were given a highlight of how fabric problems are addressed in the UK from using the FAST system. I wonder if your experience in other countries were parallel to those in the UK or are you measuring different things?

GRUPALLO: We operate the same way and we have the same experiences. In fact we exchange information presented to each other and we see if we can find the same solution to more or less similar problems.

THEVARATNAM: Is variation of results between laboratories statistically significant?

De BOOS: What I can say in defence of the results we have got is that the errors associated with FAST in various laboratories were as good or better than other systems in the market. There are differences in testing techniques, although we try to standardize these as much as possible, and differences in the sample, although cut from the same piece of fabric. All these differences went into the final statistical figure between labs, and those statistical differences are indeed quite small in comparison to my assessment of commercial difference.

POSTLE: Whatever measuring system you are using, if you want to reduce the error and increase the precision there is always the technique of increasing the number of tests, that of course adds to more work.

QUESTION: Are there systems in use bought by garment as well as fabric manufacturers?

De BOOS: It is roughly equal split, not quite so in all countries. The major drive towards the use of this technology comes from the garment manufacturer. However, fabric manufacturers saw potential benefits to them both in terms of their relations with their customers and in terms of their own process optimization and they soon adapted this technology as well. In the UK it is roughly equal split in Italy there are possibly more fabric manufacturers than garment manufacturers that are using FAST.

FAN: With regard to the FAST system, recognizing that it has been designed so that it is simple and self-sufficient for garment manufacturing problems, have you got any supporting data to show that the number of tests are sufficient for characterizing fabric problems?

De BOOS: In designing the FAST system we obviously considered the vast experience of all people working in that area before and we have chosen properties not only from our experience alone but from combining 30 years of research in garment manufacture. All I can say in defence of the system as we have it now, is that we haven't come across a fabric in the UK that was poor and not found.

FAN: So if you have those properties measured by FAST you can predict all manufacturing problems in the UK?

De BOOS: So far, but nobody would consider this technology static or imply that we know everything, I am sure that there will be such situations in the future. I do not think that there is a manufacturer of testing equipments to claim that they measure everything.

FAN: The measurement is not important but the interpretation of the results.

De BOOS: I cannot agree with you more.

LITTLE: At the recent Tokyo JIAM show several manufacturers used KESF data to set up their equipment. Is there any similar activity with FAST and equipment suppliers? What are your plans in this area?

De BOOS: There is no similar activity associated with FAST at this stage. All I would point out is that all objective measurement systems correlate with each other with some exception in terms of measured properties. So I would anticipate that if you were to use any objective measurement system to set up sewing machines, then any other objective measurement system would do a similar job.

Introduction to Professor Keith Slater

Our next speaker is Professor Keith Slater from the University of Guelph in Canada. Up to this stage we have been concentrating upon mechancial properties. Of course there is a whole range of other physical properties that relate to comfort in relation to fabric. Keith Slater was educated at the University of Leeds and he is now Professor of Textile Science at the University of Guelph. He has written a number of papers in textile and clothing areas and his paper today is 'The relation between fibre properties and fabric comfort behaviour'.

The relation between fibre properties and fabric comfort behaviour

K. Slater
School of Engineering, University of Guelph, Ontario, Canada

ABSTRACT

Comfort is a property which can never be quantitatively defined, simply because there are so many different variables involved and each of them is subjective in nature to some extent. Thus, one person may be perfectly *comfortable* in a given situation while another person at the same time and place may be very *uncomfortable*.

There are, though, some comfort states which virtually everybody can be expected to find acceptable and which are controlled to a large degree by the properties of the fibres and fabrics from which garments involved in establishing the *comfort level* are made. These include such matters as thermal retention or *transmission, moisture vapour permeability, water resistance* and *psychological comfort* factors governed by *colour, style* or *fit*.

Thus, *fibre content, yarn* or *fabric construction techniques, finishing treatments* and *garment design* are all significantly related to comfort and must all be taken into account in producing suitable apparel items. In this paper, the objective methods of measuring these parameters are described briefly, and the potential effects on comfort are discussed.

1. THE MEANING OF COMFORT

Comfort may be defined as a state of pleasant psychological, physiological and physical harmony between a human being and the environment. All three aspects are equally important, since it is impossible for comfort to be enjoyed if any one of them is absent. The presence of embarrassment, or the need to remain in a room which is too hot, or the effect of an abrasive surface on the body exemplify each type of comfort respectively, and it is easy to see how any of the three types may interfere with total comfort. The effect of the environment is also important. A human being who is saturated with water may feel perfectly comfortable in a swimming pool, but will not be comfortable if the wetting results from a storm on the way home.

A crucial point to remember is the fact that comfort is highly subjective. Two people in exactly the same environment may differ radically in their assessment of the comfort of the situation, for any of a number of reasons. Different physiological responses to the environment, different social norms or the effects of extraneous factors (such as family, health or financial worries) may be present to distort the overall comfort level of each one. Thus, it is impossible to establish any system designed to measure total comfort and extremely difficult to assess even one aspect of the subject. Not only will many of these idiosyncrasies be ignored by the investigator (or even be unrecognized by the subject), but they are impossible to control. In addition, the act of questioning a subject about comfort level can distort that level merely by the power of suggestion. All that can be done, then, is to attempt to isolate one specific aspect of comfort and devise a system of assessment which aims to eliminate, as far as possible, these extraneous effects by carrying out tests on a very large number of subjects. The tests may be objective or subjective in nature, depending on the result sought, but it must always be remembered that there may be external influences which can bias either type of test in an unpredictable manner. The effect of using textile materials is one such influence, and its importance in affecting comfort cannot be overemphasized.

2. STATES OF COMFORT

In normal experience, there are a number of comfort factors which have recognizable ranges in which a human being prefers to remain. Thus, in the area of physiological comfort, for example, a person aims to be warm enough so that the shivering mechanism is not triggered, but not so hot that sensible perspiration begins. In the same way, the body should feel dry (that is, not be perspiring noticably) but not be so dry that the skin feels itchy or begins to flake.

Physically, discomfort will typically be experienced if the body is exposed to a surface which subjects it to frictional contact outside a tolerable range. If friction is high, or of the external surface contacted is too stiff, then the skin may rub, subsequently causing severe pain and possible an open sore to develop. Conversely, if the contacted surface is too soft, a tickling sensation may occur, bringing with it discomfort of a different type.

Physiological cqmfort may be influenced by the appearance of the environment, including the appearance of any clothing or other textile article in the vicinity. Thus, the way in which a garment drapes over the body, or the colour of a clothing article (especilly where different colours must be coordinated) will influence heavily the comfort of the wearer. So, too, will the way in which it creases, or looses its crease, if the wearer is sensitive to such factors or if the article thereby becomes inappropriate for a specific social occasion. Not only the appearance of one's own clothing affects comfort; the presence of an unsuitable colour harmony in someone else can also, in many cases, adversely influence one's own feelings about the overall level of personal comfort prevailing in the situation.

3. PHYSICAL FACTORS

We can now consider the physical factors which may influence each of these comfort states. Body warmth, for instance, is a function of the external air temperature and the thermal resistance of the clothing. In summer, when air temperatures are high, garments which allow heat to escape readily are desirable and, since the body's principle mechanisms for losing heat are the processes of convection and perspiration, the garment (and the fabric from which it is made) must be ventilated for good convective heat flow and permit moisture vapour to escape. An open structure, with adequate potential bellows effect, is thus needed. In winter, however, the situation is somewhat different. In cold climates, there is a need to conserve heat, a process achieved by providing plenty of insulating 'dead air' spaces, which may be obtained, at least in theory, by preventing the escape of all the air from within a garment. The simplest way to accomplish this aim is to enclose the body in an impermeable container but, apart from the risk of suffocation if this process is carried out too conscientiously, there is also the problem of allowing the moisture vapour generated by perspiration to escape. If this is not done, there will be a rapid build-up of moisture vapour inside the garment, and consensation will subsequently take place to impart uncomfortable wetness to the wearer.

This occurrence will then produce additional adverse reactions in two major ways. In the first case, the liquid water will draw latent heat from the body in an effort to evaporate and, secondly, saturation of the garment will eliminate the dead air spaces, thus destroying the thermal insulation. The resulting drain on the heat reserve of the wearer will not only be extremely uncomfortable, but may also in the limit produce frostbite or other serious complications, including the limit produce frostbite or other serious complications, including death from exposure.

It is important also to recognize the specific dangers associated with working in a very cold climate. If hard work is in progress, the body will produce a lot of heat but, when a rest period is taken, this heat production will cease. The production of perspiration for cooling purposes, however, will tend to continue for some time, so that an excessive amount of moisture is being generated and, with no heat production to help it evaporate, there will be a rapid condensation, leading to the severe chilling effect just described. Under these conditions, then, it is important to provide some means of adjustment (such as a neck closure or belt) to control the convective cooling and the escape of perspiration in tandem so that thermal and moisture movement can take place in the optimum manner for comfort retention.

This in turn brings to mind a similar type of problem, that of maintaining dryness in wet weather. Again, the ideal solution is to enclose the body in an impermeable layer through which liquid water cannot penetrate, but the escape of perspiration moisture vapour is once more prevented by this action. In this case the incorporation of a neck opening is pointless, since this would provide an undesirable entry point for the water. The dilemma is not easy to solve, and usually involves providing double layer structure to prevent water pressure driving rain into the garment, together with a water-resistant finish to increase surface tension, thus encouraging liquid water to bead while allowing moisture vapour to escape through the relatively open pores of the cloth.

Physical comfort, in which abrasion is the crucial factor, is not often a problem with textile materials themselves, since they normally have frictional properties which fall into an acceptable range for this purpose. The exceptions (such as the tubing taking place at the heel of a stocking) usually involve some other item of clothing (the shoe in this case) which is exerting some force on the textile and causing it to move repeatedly across the surface of the skin, thus producing a blister or other type of open sore. Some kind of cushioning effect, supplied either by adding bulk to the fabric or by adjusting the elasticity of the structure to suit the conditions of movement, is often incorporated at such contact positions to prevent this type of damage.

Psychological factors of importance to comfort must be controlled by careful attention to the aesthetic aspects of the clothing system. Thus, the colour and design of the fabric from which a garment is made should be selected with a specific end-use in mind. The style and fit, too, should match the personality of the wearer, and the entire clothing system should be coordinated to ensure that a pleasing total effect results from the combintion of colours and styles selected if fashionability is of prime concern.

4. THE MEASUREMENT OF COMFORT-RELATED PROPERTIES

As mentioned already, it is not possible to measure comfort as a single entity. Each of the factors contributing to the total comfort level can, however, be assessed individually, though it is not always possible to relate the assessment directly to a specific comfort state. The major difficulty is the matter of subjectivity, since the opinions of different people as to what constitutes acceptable comfort may be incompatible. In addition, personal idiosyncrasies, cultural biases, the effect of climate conditioning and the mental state of a subject can all influence the replies to questions about comfort. Consequently, the assumptions made by analysing the replies may well be biased in an unknown and logically unpredicatable manner, so that the results assumed may be erroneous.

Objective tests, on the other hand, do not necessarily give reliable estimates of the comfort factor that they are trying to measure. The reason is that the laboratory test undertaken may not simulate exactly the practical situation which is being assessed, or there may be added factors which the laboratory test fails to take into account. Thus, a fabric which would apparently be excellent for excluding rain, from the results of its water resistance tests, would not necessarily be comfortable in perspiration conditions, as discussed above, because the water resistance test fails to include a component to allow for the production of internal moisture by perspiration.

With reservations of this kind in mind, then, it is possible to descibe the methods used to estimate the comfort potential of textile materials. Thermal resistance, for instance, is assessed by placing the textile in series with a heat source, a heat sink, a standard heat flow barrier of known thermal resistance, and a heat meter, so that the heat flow rate through the apparatus can be established. Measurement of the temperature at various sites in the system then allows an estimate to be made of the unknown thermal resistance of the fabric. The technique is applicable to a layered structure, so that an entire clothing assembly can be evaluated as a single unit if this is desirable for Arctic wear or in similar situations.

Liquid water resistance is measured by exposing the fabric to a source of water under pressure, either in a hydrostatic test or in a dynamic one in which simulated rain is allowed to impinge on it. The results may be assessed visually in a empirical way, or an attempt may be made to obtain quantitative measures by weighing or otherwise evaluating the amount of water penetrating the cloth. Experts agree that none of the laboratory tests is capable of giving exact prediction of behaviour in rain, but most of them can give reasonably reliable estimates of how two or more fabrics compare in relative terms with each other. Again, the test can be carried out on a multilayered structure, so that the effectiveness of an entire clothing system can be assessed if so desired.

Moisture vapour transmission is not as easy to investigate. Difficulty arises because moisture vapour and air molecules are comparable in size and, since air is so omnipresent, it tends to complicate the measurement. The technique most often used is the rotating dish one, in which a comparison is attempted of the evaporation taking place from one set of dishes covered by a standard fabric and the one unknown moisture vapour resistance. The test is, however, very tedious to carry out and the results obtained from it are viewed with suspicion by most workers. A different approach involves the establishment of a moisture vapour gradient, in which the textile fabric is placed to interfere with the flow of the vapour and the impedance of its presence is obtained by relative humidity sensing. This method (unlike the dish one) can be used for a clothing assembly, but has not yet been investigated rigorously or received widespread acceptance.

Potential abrasion effects may best be assessed by estimating the coefficient of friction, using any of a number of techniques which all depend on the stick–slip principle (a differential method which alternates between static and dynamic states to enable the coefficients of starting and sliding friction to be determined together in a single experiment) for their operation. By its very nature, such a test is restricted to a single surface, which makes it difficult to interpret the meaning of laboratory results in a practical situation. In addition, of course, the tendency of a fabric to abrade into a hole can be evaluated by using one of the standard abrasion resistance testing machines to give some indication of how effective the fabric is as a barrier in preventing such wear from removing the protective layer of textile from the skin. These machines subject fabric test specimens to reproducible abrasion against standard surfaces, of metal or of an abrasive grit compound, in order to determine how resistant each one is to the effect of such abrasion. Again, test results can not always be correlated precisely with practical situations, but useful comparative values may normally be obtained with most of the machines.

Finally, some of the psychological comfort factors can be mentioned. The colour of an article, or its closeness to a desired shade, can be assessed subjectively, but a more precise approach is to use a colour difference meter. This device illuminates the surface of the cloth with white light and measures the colour of the reflected rays to determine how their contact with the fabric has changed the light/dark, green/red and yellow/blue components of light. A numerical value between zero and one for the first of these and in the range plus one to minus one for the other two, is obtained, so enabling quantitatively comparisons of different fabrics to be made very simply.

In the matter of fit or style, it is impossible to be precise about tests. Personal taste, figure type, age, range of activity intended and similar considerations must be taken into account when deciding whether to aim for a 'tight' or 'loose' fit. In addition, factors such as health or self-image can govern what is meant by these two terms, since they may be interpreted differently by different people. All that can be done is to recommend the perusal of fashion magazines, in conjunction with a knowledge of individual preferences, in order to select the article which is most flattering or pleasing to each specific individual.

5. THE EFFECTS OF TEXTILE PROPERTIES

Finally, the way in which all these desirable or undesirable attributes can be affected by the actual textile construction should be considered. Fibre, yarn or fabric parameters, together with finishing treatments, can all exert considerable influence on the ultimate physical properties and hence on the comfort level produced by a garment.

The three most crucial fibre properties are type, length and fineness. The fibre content determines to a considerable degree the durability (in terms of strength and abrasion resistance, for example) of the end-product, hence governing not only the tendency to disintegrate into an unsightly hole but also the mental comfort of obtaining an acceptable article life time for the price paid. The fibre content factor usually regarded as paramount, though, is the one making the greatest contribution to comfort, the absorbency. This exerts an effect in turn on the thermal balance, the moisture vapour or liquid water permeability, the electrostatic propensity, and the launderability, all of which have obvious implications for mental or physical comfort. Frictional effects are also governed to some extent by fibre type, again affecting comfort.

Fibre length is important in determining the strength, lustre, evenness and hairiness of the yarn, and hence various comfort attributes of the fabric and garment. A long fibre will, in general, produce a stronger yarn than a short one, since twist will allow greater frictional contact to maintain the structure against tensile stress. Long fibres also tend to lie more parallel to one another, so that a more lustrous and even yarn, with less hairiness as a consequence of fibres oriented at a large angle to the yarn principal axis, results.

Fibre fineness influences bending length, since a fine fibre will have less resistance to transverse deformation than a coarse one. Thus, garments made from finer fibres will tend to have better drape (thus being aesthetically more pleasing) and to be softer (with consequent beneficial effects on frictional skin contact) than those made from coarse ones.

Yarn properties, governed to some extent by fibre ones, are also crucial in affecting comfort behaviour. A fine yarn, like a fine fibre, will tend to be more flexible than a coarse one, and may be achieved without loss of tensile strength by the use of longer fibres. The finer yarn will, naturally, be weaker if fibre length is the same, so care should be taken to use long enough fibres to compensate for the lower number of frictionally contacting surfaces in the cross-section. High yarn count also increases thickness and weight, both of which bring about resulting changes in the warmth and loading effect of a fabric.

High yarn twist leads to a stronger yarn, but may also cause unwanted frictional characteristics because the yarn is 'harder' and, in addition, a taught twist tends to make the yarn stiffer, reducing drape and softness in the fabric. Tight twist, though (along with longer fibres) improves yarn evenness, thus enhancing strength and lustre, but reduces bulk (and hence thermal resistance) and cover (the percentage of the fabric which is a fibre-filled, rather than open air space, area) in a fabric.

Critical fabric properties include this cover also influenced by the spacing of yarns), which has a drastic effect on air or moisture vapour permeability and also on thermal resistance. If cover is high, there are fewer pores for the passage of gases, but there is also a better chance of retaining dead air within the structure. Thickness, governed (as implied above) by yarn spacing, twist or count and weaving tension, also modifies these same properties and exterts a major influence on weight. This is reflected in the load carried by the wearer of a garment, so that the yarn properties and other fabric construction conditions can be directly responsible for fatigue as a result of this loading factor. The choice of a woven or knitted structure, and the variables selected for either one of these methods of manufacture, will also have an effect on the cover, thickness and weight of the fabric. In general, knitted structures are more elastic then woven ones, and are more permeable to the movement of water vapour or air, so making them easily shaped and more comfortable in warm conditions. Conversely, however, a woven fabric will tend to keep its shape better when subjected to deforming forces and will retain heat better in winter. Cloth bulk (a measure of the air/fibre ratio) is dependent, too, on many of the properties already mentioned, and has notable influence on thermal resistance, moisture vapour transmission and air permeability. This property, from which air entrapment to change thermal resistance and permeability to air or moisture vapour results, is highly dependent on individual construction factors; it is increased by loose twist, yarn unevenness or hairiness, high yarn count and shorter fibres, for instance.

The last area to be considered is that of finishing treatments after the fabric is manufactured. Finishes which impart specific changes likely to affect the comfort level of a fabric may be mechanical ones, such as brushing (which increases bulk) or shearing (with the opposite effect), or they may be chemical ones. Of these, dyeing and printing are the most obvious ones, with their appreciable effect on visual appearance (and thus on psychological comfort), though other chemical finishes, such as those intended to enhance crease-resistance, stain resistance, antistatic performance, flame resistance or water resistance clearly have a vital role to play in enhancing comfort of a psychological or physical kind or in improving safety.

After the fabric has received all the necessary treatments for optimum performance in its intended end-use, of course, there are still the garment construction techniques, such as tailoring, to be used in the enhancement of comfort, but these are, in general, outside the topic of this paper. It should be noted, however, that major differences in convective cooling, heat retention, air or moisture vapour transmission and liquid water resistance can be achieved by careful design of clothing and by selection of appropriate combinations of cloth. So, too, can aesthetic factors, all these being additional benefits superimposed on the obvious one of enhanced psychological comfort.

6. CONCLUSIONS

The subject of comfort is a complex one, highly influenced by personal opinions, idiosyncrasies and background. It is impossible to analyse comfort as a total property, though some attempt may be made to examine individual factors making a contribution to its level. This examination may include both subjective and objective factors, though it should be realized that the results obtained by it may be at risk of biases which are both unpredictable and unidentifiable.

In general, though, textile properties are crucially important in establishing the level of comfort of a finished article (usually a garment) for the end-user. The fibre, yarn or fabric parameters, including the construction conditions used, and the finishing treatments applied after the cloth is made, can all exert considerable influence on the behaviour of the article, and subsequent operations (notably those of garment manufacture) can add further modifications in many cases, particularly where psychological comfort aspects are concerned.

For satisfactory performance, then, it is necessary to consider very carefully the conditions under which the article is to be used, so that the optimum combination of psychological and physical comfort can be achieved in the environment under which use is envisaged. The textile manufacturer should therefore take into consideration all the user's requirements and the specific end-use envisaged before attempting to produce goods designed to enhance comfort.

QUESTIONS

FAN: You have spoken about fibre, yarn properties, psychology and energy factors related to comfort. What about the fabric properties related to the fitting of garments which is also important to comfort. I think that this may be even more important than fabric properties such as heat and moisture resistance of the fabric, because three-quarters of heat or moisture go through the opening of the garment not through the fabric. You have discussed about heat and moisture transfer through the fabric, but I thought that perhaps those fabric properties related to garment fit of the body are more important.

SLATER: Several comments: first of all the subject you describe is vast; secondly there are people here far more knowledgeable than I am in what you are asking; and thirdly the number of properties related to fitting come under tailoring which is coming up tomorrow and the day after.

THEVARATNAM: Do you see any development in predicting comfort in the future?

SLATER: No unless you start dictating what people's comfort would be. You will be itchy or hot at this temperature. Unless you have that kind of situation I do not see it, people's idiosyncrasies are so different. I am now become acclimatized to the Canadian climate which is much colder than here, because anything above 4°C is regarded as hot in Canada. So you have the personal idiosyncrasies, you have acclimatization, you have individual variation in metabolism and so on. I really do not see how you can predict comfort like the handle of the fabric as total, you may, however, isolate various aspects of comfort.

DeARAUJO: What about if you have a fingertrip or snakechart of people?

SLATER: This has been tried. The variation is so high that it does not seem to work. We have to take into account the whole population.

SHISHOO: In the last decade or so tremendous progress has been made to develop objective measurement methods to correlate with what happens in practice, it is all difficult, however, one should talk about the latest developments on objective measurement of comfort properties not the final prediction of comfort. We are very near to finding an appropriate method to measure fabric properties, to measure properties of fabric layers, and to measure garment properties. It is appropriate to say that research in the last decade or so has come closer to finding such an objective measurement testing method.

SLATER: I would say that there has been little progress and I am not as optimistic as you are.

Session 2: Application of textile objective measurement to improve sewing performance of garments

CHAIR: Professor J. C. Furniss
Philadelphia College of Textiles and Science, USA

INTRODUCTION

The opening remarks made yesterday gave you the background to this conference, to the importance and implications of textile objective measurement to the garment industry today and in the future; but today we are turning our attention to the application of textile objective measurement to improve the sewing performance of garments. As already mentioned yesterday, the objective measurement of textile fabric properties in order to predict their sewing performance and garment quality will become increasingly important as the international competition heats up in the world apparel markets. Although receptive to Japanese apparel manufacturers, objective measurement has been much slower in the USA and Europe. To my knowledge no USA apparel manufacturer has purchased the KESF equipment set; the present sets are mainly in universities and colleges with textile schools and departments. A few non-textile manufacturers, interestingly enough, who are in mechanical fields have purchased the equipment. I think that the high cost of the KES, and the high skill required to operate it, have been used by manufacturers as the main reason for not becoming involved. However, I know that in colleges and universities in the USA we have constant requests to do tests for the industry. I think now we see in the market place cheaper possibilities like the FAST and possibly others that we may hear about today. Then industry may respond more positively, because they mainly look at it as a means of quality control and I think that the KES system is seen very much as a research piece of equipment. However, having said that, the objective measurement for predicting fabric performance is the key for successful introduction of future robotized, or maybe one day fully apparel automated systems. I hope that we are going to learn quite a lot from certainly one expert in the field whom I would now like to introduce to you.

Introduction to Professor Sueo Kawabata

Our first speaker today, is Profesor Sueo Kawabata who is known to us all as the major worker in this field. Professor Kawabata is in the Department of Polymer Chemistry at Kyoto University. In 1960 he graduated from the graduate school of the University's department of polymer chemistry and received a doctorate of engineering in 1961. After graduation he became a research associate in the department and his research field was expanded to a wider range covering textile technology. In 1964 he spent a year in USA as a research fellow in materials engineering. After returning to Kyoto University in 1965 he became associate professor and later in 1973 he received a chair in the department of polymer chemistry. His research field is divided into two areas: the first area is in textile technology and textile fibre mechanics, and the second area is in mechanics of polymer solids. He initiated his research on the objective evaluation of fabric handle on the basis of his fundamental research in fabric mechanics around 1970, because of his impetus to improve textile quality by means of modern engineering. He received the first memorial award from the Textile Machinery Society of Japan in 1971 for his research into textile technology, and also received a scientific award from the Society of Polymer Science in Japan in 1974 for his work on the finite deformation of solids. In 1987 he was awarded a memorial medal for his scientific activities in textile technology from the Textile Institute in the UK, and more recently, in 1989, he received an International Science Prize for his work on the objective measurement of fabric handle. The title of Professor Kawabata's paper is 'Recent progress in the application of objective measurement to clothing manufacture'; along with him are: Professor M. Niwa, K. Ito and M. Nitta.

Recent progress in the application of objective measurement to clothing manufacture

S. Kawabata
Department of Polymer Chemistry, Kyoto University, Japan
Masako Niwa
Department of Clothing Science, Nara Women's University, Japan
K. Ito
Osaka Apparel Technology Modernization Association, Hirakata City, Osaka, Japan
M. Nitta
Oga Clothing Hirakata Co., Hirakata City, Osaka, Japan

ABSTRACT

The application of objective measurement of the mechanical properties of fabrics in the apparel industry began around 1975 in the Hirakata area, which is one of the centres of men's suit production in Japan. At that time the KESF system was just developed and hence spread rapidly. The measurement of *mechanical data under low-load level* by the KESF provided useful information for the *apparel engineers* who needed some means of *fabric measurement*, by which the tailoring process might be controlled. The fabric *dimensional stability* testing using *steam press* was also standardized at that time (HESC 103A Method).

At present, the KESF data and the stability data are essential for apparel engineers and are used widely in the *Japanese apparel industry*. In addition to the use of objective measurements in each factory, a centre for *objective fabric inspection* has been recently initiated in the Hirakata area, for the *inspection and control of fabric* by the *objective system for tailoring process control*.

In addition, a cooperative work between the apparel engineers and the University has been carried out to develop a new equation for *predicting good appearance* of a suit on the basis of fabric mechanical data. Automatic tailoring such as *automatic over-feed action* on the basis of fabric mechanical property is also carried out under the cooperation of University/apparel industry/sewing machine manufacturer (Juki) in Hirakata. The progress of those projects will be presented.

1. INTRODUCTION

Fabric property in relation to 'better fitting to human body' is an essential requirement of clothing material. This important property of fabric has been evaluated subjectively by hand touching called handle judgement by not only consumers but also professional textile engineers. Because of the importance of the engineering design of fabric quality concerning this fabric handle, the objective method of evaluating the fabric handle was developed in 1972 by Kawabata and Niwa with the co-operation of the experts of handle judgement in textile industry in Japan through the Hand Evaluation and Standardization Committee (HESC) organized by Kawabata [1]. This objective method started its spread to industries around 1975 and has provided us with a powerful tool for developing new quality fabrics. The basic mechanical properties and surface property of fabric are measured under low load condition. The load applied on the fabric tested is near the load when fingers pull a fabric or deform it at the handle judgment.

In parallel to the spread of the use of this objective handle evaluation in the textile industry in Japan, the apparel engineers who were members of the HESC were interested in the direct use of the fabric mechanical parameters, which were used for the objective handle evaluation, for process control and fabric buying around 1978 with the cooperative technical research with universities [2,3]. This is because the men's suit manufacturing in Japan had shifted its structure from the past tailor shop style to industrial structure. And along with this shifting, the apparel engineers in the industry have faced many difficulties in the quality control of their products around 1975, in order to prevent 'defect products'. Researches for solving this problem were carried out with the cooperation of the apparel engineers and the universities through the activities of HESC in 1978 and one of the conclusions obtained at that time was that fabric tensile, shear and bending properties are particularly important for suit manufacturing. This conclusion was made based on the analysis of the fabric data accumulated in two or three factories in the period of several years. Then the joint research begun for investigating methods of preventing defect products.

Another conclusion obtained at that time is that the fabric press-shrinkage caused by the steam-press processes on suit production line is also one reason of producing defect suits, and the fabric property corresponding to this shrinkage and the recovery property must be measured precisely to prevent the defective appearance of the suit produced. This shrinkage and recovedry property is a dynamic property and the testing method named HESC-CFT-103A method [4,5] has been designed to measure the dynamic property of fabric shrinkage and recovery and completed in 1979 by the HESC. The original idea of this method had been already in practice in a factory by Ito. The systematic investigation to standardize the steam-press shrinkage test was carried out by the HESC members from suit apparel and finishing companies. Among the many trial methods inspected in this project, the '103A' method was selected as the standard test method. This test procedure was soon employed by the HESC member factories. More than ten years have passed since then, and this '103A' method has settled in with some suit manufacturers. During the practical use of this test method improved procedures of testing have been developed in every factory for easier operation of the measurement under the condition that the parameters measured by these modified methods must be the same as those in the original method. Testing has been effectively applied to process control in the production line of a suit manufacturing company [6].

The manufacturing of non-defective suits was the main target in the engineering of suit manufacturing using line production systems, and this has been achieved by using objective measurement of fabric. Now, the engineering is going towards the direction of the more 'high quality suit' manufacturing on the basis of fabric objective measurement. Niwa and Nitta have continued the inspection of the reliability of the equation predicting the suit appearance on the basis of fabric mechanical property [7,8]. This research was initiated by them and Kawabata around 1978 aiming at the production of better quality suits. And equation 'KN eq. 10' and 'KN eq. 10LC' are now almost standard [9,10]. Equation 'KN eq. 10' can be applied to both summer and winter suitings commonly except linen or linen blended suitings to which the 'KN eq. 10LC' is applied.

Another direction of recent researches in our group is the arrangement for the design of automated apparel production system from a basic concept. One of the most difficult but essential subjects is the design of an automatic over-feed sewing machine. The joint research with Juki-Co, which is a member of HESC, is now going on to investigate the interrelation between sewing and fabric mechanical properties.

The KESF instrument for the fabric mechanical testing under low load level has been designed for the purpose of the objective evaluation of fabric handle originally and also as a testing system for laboratory research [1,2]. Now, the use of this instrument has been spread not only to textile industries but also to a wide range of industries. In the textile industries, the instrument is now used on production line of garments. In order to meet this situation, a new KESF is being developed. The instrument measures exactly the same parameters as those measured by the original instrument of the KESF, but is operated completely automatically at high speed. The details about the recent progress in the researches on apparel engineering is introduced in this presentation.

2. THE SPREAD OF THE PROCESS CONTROL ON THE BASIS OF FABRIC MECHANICAL PROPERTIES IN APPAREL MANUFACTURING

In 1983, Ito and Kawabata presented a chart for the process control of men's suit production [11]. This chart which is shown in Fig. 1 indicates the fabric properties related with difficulties during processing using tensile and shear properties of the processed fabric. If these properties of fabric fall in the 'non-control' zone in Fig. 1, any particular notice is not required for the workers on the production line to take extra actions for this fabric and a non-defect suit will be produced without any fabric trouble. On the other hand, if some of the properties of a fabric fall outside of the zone, some extra and appropriate actions of the workers are required accordingly following the guidelines in a manual. This control system is now fully applied at major men's suit companies with some modifications made by some companies based on their own database.

The hatched area shown in Fig. 1 is the zone in which the mechanical properties of excellent fabrics for high quality suit fall. The suits made by these fabrics are

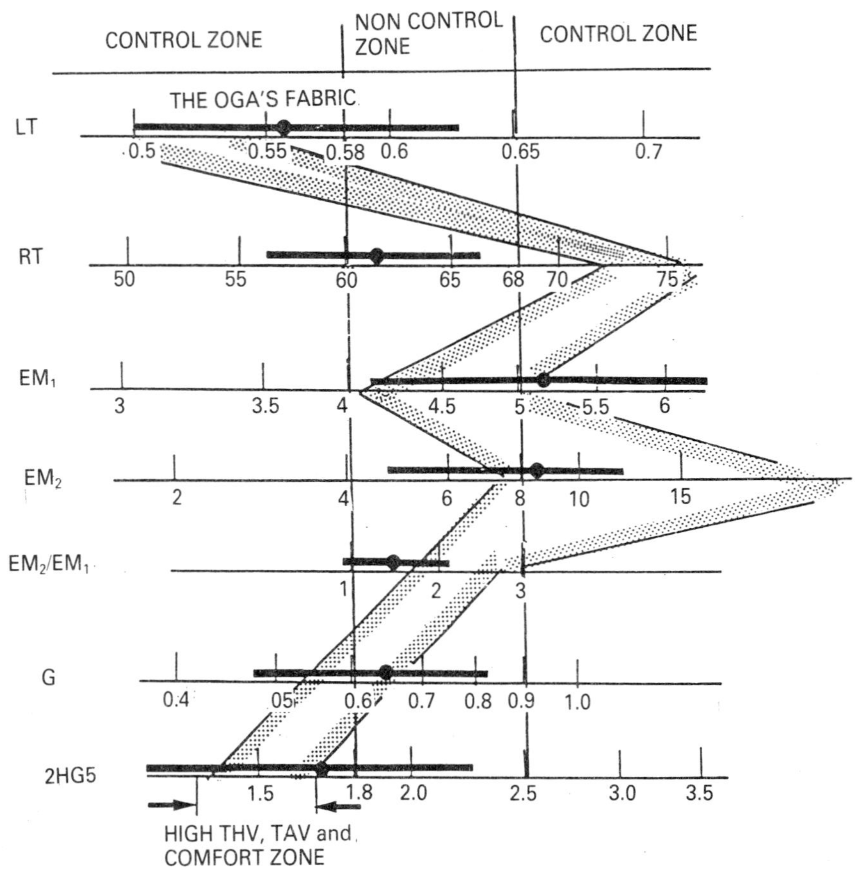

Fig. 1 — Black dots show the mean of the mechanical parameter of the fabrics which were tailored in 1989 on the production line in Oga Clothing Hirakata factory, they tailored approximately 400 kinds of fabrics in the year. Horizontal broad line shows its standard deviation. This shows a typical trend in recent apparel industry, that is more high quality suit production and fabric process control is necessary.

predicted to be comfortable in wear and beautiful in its appearance. It is noted, however, that the mechanical properties of these high quality fabrics are mostly out of the 'non-control zone'. This suggests to us that a high quality suit is made by such a high quality fabric, however, its tailoring process in production is difficult. In such cases, the process control on the production line is required. This process control was firstly introduced by Ito at Sanko-Iryo Co. on the basis of the KESF mechanical data and fabric press-shrinkage property in 1978 and he reported the results in 1983 [6]. Since then, such process control based on fabric properties has become common in men's suit manufacturers in Japan. Every company has the control system despite the fact that they are slightly different to each other because of the different policy of production engineering and the different layout of production [12].

The percentage of the fabrics which fall in the control zones is shown in Table 1. The data in this table is based on the statistics of the measurements of 6300 kinds of

Table 1 — Percentage of the fabrics which fall in the inside or outside of the non-control zone. From Ito's statistics based on the fabrics processed mostly in Hirakata, Osaka area in the period from 1988 to 1989 (unit: %)

	Fabrics in non control zone	Fully or partially in control zone		
		Rank A[a]	Rank B[b]	Rank C[c]
(1988)				
Fall-winter	15	20	55	10
Spring/summmer	5	5	65	25
(1989)				
Fall/winter	15	20	55	10
Spring/summer	5	5	63	27
Very high quality fabrics				
(1988–1989)[d]				
Fall/winter	5	70	25	0
Spring/summer	3	30	57	10

[a]Rank A: mostly within the high quality zone. Good suit will be manufactured by the sewing process control.
[b]Rank B: some properties are outside of the high quality zone. In particular, those fabrics having too small EM1 and/or EM2 included in this rank are difficult to control even though the special notice for the control is sent to the workers on the production line but the other fabrics of this rank will be possible to control to make a good suit.
[c]Rank C: the properties are mostly out of the control zone to poorer direction. The control is difficult and a good appearance and comfortable suit cannot be manufactured by these fabrics. These fabrics have particularly low extensibility and high shear modulus and hysteresis.
[d]Samples: the winter suits are almost all wool fabrics and summer suits are wool worsted fabrics; 50%, wool/PET blend fabrics; 35% and linen, silk blend fabrics; 15%.

fabrics processed in the men's apparel companies in the Hirakata area of Osaka including some from other parts of Japan in a year.

In Fig. 1, a black dot and a range for each mechanical parameter show the mean value of the mechanical properties of the fabrics processed at a factory of Oga Clothing Co. in 1989 and the standard deviation respectively. As seen from this figure, the large part of the fabrics are out of the non-control zone. Table 1 and Fig. 1 show a recent trend, that is, higher quality fabrics for better suits rather than the suiting fabric having easier processability. This is a general trend recently in Japan and the reason of the spread of the control system in apparel manufacturers.

The press-shrinkage and recovery behaviour of the processed fabric including relaxation shrinkage behaviour are important factors also to eliminate imperfect suit appearance. The parameters measured in the HESC-FT-103A method standardized in 1978 are shown in Table 2.

To reduce the time for testing, the fabric dimension at the equilibrium state is estimated from the water content of fabric by measuring the fabric weight at the earlier time than the time at equilibrium. This method has been used in practice

Table 2 — The dynamic fabric-shrinkage test. Parameters measured by the HESC-FT-103A method [4,5]

Parameter	Definition (fabric shrinkage is defined by the shrinkage referring to the size of the fabric in the stage before steam-pressing)
S_1	fabric shrinkage immediately after the steam-press
S_2	fabric shrinkage at the equilibrium-state in room condition after the steam press
S_3	fabric shrinkage immediately after the humidification in 100% RH condition, the S_3 takes a negative value usually, that is, expansion value. S_3 gives the maximum hygral expansion value.
S_4	fabric shrinkage measured in the equilibrium state in room condition after the relaxation process with water. The S_4 gives the estimated value of the relaxation shrinkage of the fabric.
Q	Rate of steam press-shrinkage recovery, estimated by the dehydration recovery at 2 hours after the steam pressing.

by several factories for more than 10 years and every company has modified the testing procedure for getting S_i and Q values during these ten years. The procedure which has been used at the testing and sponging centre of the Osaka Apparel Technical Modernization Association (OATMA) is shown in Table 3. This method is one of the simplified methods which has been prepared by the HESC. The value Q, which is the rate of the recovery of press-shrinkage after steam-pressing, is also estimated by the rate of dehydration of the fabric after the steam pressing by measuring the fabric weight at immediately after the steam-pressing and 2 hours after the steam-pressing. These parameters are now used for the process control of suit production line together with the mechanical properties measured with KESF. For example, S_2–S_4 provides the danger for 'bubbling' which is the wave formation on the surface of suits during wear. The OATMA looks at S_2 and S_4 in particular to decides whether a fabric is ready or not for tailoring processing as shown in Fig. 2.

On the suit production line, fabrics are applied to steam-pressing frequently. In order to accelerate the recovery of the fabric shrinkage caused by steam-pressing, steam chambers are used. The humidity in this chamber is kept at nearly 100% RH. After steam-pressing of a fabric, the fabric is placed in a steam chamber for a few minutes. This steaming is necessary for preventing seam puckering. The time for steaming must be controlled by KESF fabric mechanical data and the steam-press shrinkage data, when the Q value of a fabric is small, longer time of steaming is required. The system used in Sanko Iryo Co., Osaka adopts a batch processing system and Melbo Clothing Co., Osaka uses a belt-conveyor system. The speed of the fabric going through the chamber is controlled for every piece of fabrics based on the fabric properties as in Sanko-Iryo Co., the Melbo company has a custom-made

Table 3 — HESC-FT-103A steam press shrinkage testing method, the 'simplified procedure'

Steam press; three times of the press on the Hoftman-flat bed press under the controlled condition.		
Measurement	(1) Fabric shrinkage referred to the size in the state before pressing. (2) Fabric weight for:	
S_1:	Immediately after the press.	
S_2:	After equilibrium in room condition, estimated by the shrinkage and weight ar 2 h after the pressing.	
Steaming; in the controlled box of 95% RH and 40°C for 2 h for humidification.		
S_3:	Immediately after the humidification. (= the maximum hygral expansion)	
S_4:	In equilibrium state after drying up in room conditions, estimated by the measurement after 24 h drying in room condition (= relaxation shrinkage)	
Q:	Rate of the press-shrinkage recovery, estimated by the dehydration recovery at 2 h after the pressing. $Q = \Delta W_R / \Delta W \times 100\%$, ΔW ; Weight of the water lost by the steam press, ΔW_R; Water weight absorbed for the period of 2h after the team press.	

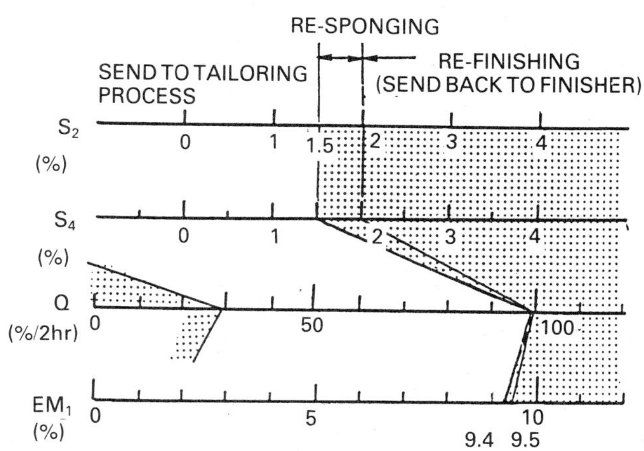

Fig. 2 — Decision chart whether fabric is ready or not for tailoring processing after sponging.

tailoring system to accomodate different kind of suit movement in the production line.

3. BASIC RESEARCHES FOR AUTOMATED SUIT MANUFACTURING SYSTEMS

We are now involved in several investigations concerning the basic problems in suit manufacturing automation of the future under the cooperation of Juki Co., Tokyo, a sewing machine maker and also a member of HESC. Two topics investigated recently are introduced as following.

3.1 The size deviation in finished suits

When a size error of a suit under the tailoring process appears for any reason during custom tailoring, it is possible to mend it to its correct designed size. The industrial line production system has, however, little opportunity to do the mending. The size of the finished suit must correspond exactly to the designed size without any mending in the production line. The research recently carried out on the size-change of suits proves the correlation between the size error of finished suits and the KESF data of the fabric.

An example of the size error occurred in the production line is shown in Fig. 3.

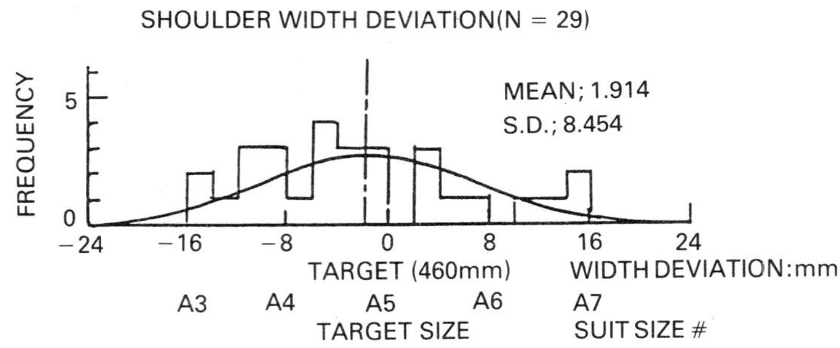

Fig. 3 — An example of the size deviation in the shoulder width of the suit during the processing on a suit production line.

The reason of this size deviation is complex. A fabric is subjected to several times of steam-press operations during the tailoring process. The relaxation shrinkage of the fabric is one of the reason, however, and almost all fabrics processed in factories are now controlling their relaxation shrinkage in the range less than 1.5% or 1% during sponging. It is noted that the size deviation tends to shift the suit to a larger size. The most possible reason is that the steam-press operation sets the fabric dimension different from its original size, mostly in larger size from the original size of the fabric. The mechanism of this size deviation is still complex, but the analysis of the data-base accumulated for those ten years shows that the fabric mechanical properties are related closely with the size deviation. The prediction equation of the size deviation has been derived by Niwa and Kawabata on the basis of the data accumulated in Oga Clothing Co. by Nitta as follows.

The prediction equation of the size deviation appeared in the back centre seam of men's suits is as following;

$$Y = C_0 + C_1\frac{X_1 - m_1}{\sigma_1} + C_2\frac{X_2 - m_2}{\sigma_2} + \ldots + C_7\frac{X_7 - m_7}{\sigma_7} \qquad (1)$$

Y is the size deviation in mm, in this case. The designed length is 700–720 mm. Mechanical parameters X_1, X_2, ... and X_7 are shown in Table 4. The fabric

Table 4 — The fabric mechanical parameters X_i

X_i	Mechanical parameter	Definition	Remarks
Deformation property:			
X_1	Extensibility	$= \log_{10}(EM/LT)$	$= X_{11}{}^a$
X_2	Effective bending stiffness	$= \log_{10}\{M'(1) + 2HB/2\}$, $M'(1) = B.K$, $K = 1$	$= X_{12}{}^a$
X_3	Effective shear stiffness	$= \log_{10}\{Fs'(1) + 2HG5/2\}$, $Fs'(1) = G.\phi$, $\phi = 1$	$= X_{13}{}^a$
Elastic property:			
X_4	Tensile resilience	$= RT$	
X_5	Bending elastic potential per unit area at $k_m = 2.5\ cm^{-1}$	$= \log_{10}\{B(k_m - 2HB/2B)^2/2\}$, $k_m = 2.5$	$= X_{21}{}^a$
X_6	Shear elastic potential per unit area at $\phi_m = 8°$	$= \log_{10}\{G_R(\phi_m - 2HG/2G_R)^2/2\}$, $G_R = G + (2HG - 2HG5)/5$	$= X_{22}{}^a$
Fabric weight			
X_7	Fabric weight per unit area	$= W$	

K, Bending curvature (cm^{-1})
ϕ, shear angle (degree)
M', Elastic bending moment ($gf.cm.cm^{-1}$)
Fs', Elastic shear force ($gf.cm^{-1}$)
[a] X_{11}, X_{12}, X_{13}, X_{21} and X_{22} are the parameters applied to the prediction of suit appearance [9,10]. EM, LT, B, . . are the fabric mechanical parameters defined in the KESF measurement and for units of them, see Appendix 1.

mechanical parameters X_i have been derived by combinating KESF parameters for the prediction of suit appearance and it has been found that these combined parameters lead to the highest regression than the use of the individual parameters of KESF [9,10].

This prediction equation provides a trend that the size deviation occurs in the larger size direction when a fabric has the properties which are; extensible but small resilience in extension property, stiff and larger hysteresis in bending property and heavy fabric-weight. The correlation coefficient (R) between the regressed and the experimental value of Y is 0.79 using the data of 140 suits tailored by the 28 kinds of suitings which were sampled so as to cover a wide range of mechanical property and fabric weight.

The prediction of the width of the shoulder part has been derived using the same mechanical parameters as in Table 4. The Y value in this prediction is the width size deviation in %. The designed length is in the range of 450–472 mm. This equation is shown in Table 5. The equation suggests that the deviation occurs to the larger size direction when the fabric property is extensible and inelastic in extension (in this case, these are the weft directional properties) and stiff and inelastic in both bending and shear deformation. The regression correlation (R) of the equation is 0.90. Another analysis suggests that the size deviation has a high correlation with formability F_2 [8]. The higher the value of F_2, the larger the elongational deviation.

$$KN(\text{eq.}(2)){:}\ F_2 = \frac{EM_2}{F_{\max} LT_2} B_2 \frac{G}{2HG5} \tag{2}$$

From Ito's experience on the basis of his database which consists of 3600 fabrics data accumulated in his testing centre, the size deviation in the shoulder width of finished suits is shown in Table 6. Also an example of his action for such dangerous fabrics is shown in Table 7. His experience is well predicted by the equations shown in Table 5.

In many factories at present, such a warning system is becoming popular and the data for the warning are prepared by the Testing and Sponging Centre as shown later. And each factory has their own manual for preventing the size deviation. They are, for example;

(1) Careful and strict indications for operations are given to workers in the production line for the fabrics, having predicted difficulty for the size deviation (Oga Co., Sanko Iryo Co., Melbo Co., etc.).
(2) Use of the computer program for automatic cutting and adjust the pattern size for every fabric based on the KESF data of the fabric which the customer ordered. The Melbo Co. has a custom tailoring on line production systems, where the body size of a customer is measured at a sales store then sent to the factory and the fabric is cut by an automated fabric pattern cutter for processing.

3.2 Automatic overfeed sewing machine

The development of the automatic overfeed sewing machine is essentially important for the future of automated tailoring. The authors are also involved in a joint research project on this issue with the cooperation of Juki Co. which is a sewing

Table 5 — Prediction equation (1) of size deviation appear in the back centre seam length and the shoulder width of men's suits[c] (unit mm)

Back-centre-seam length: Y_{JL}				Shoulder width: Y_{SW}			
X_i	C_i	m_i	σ_i	X_i	C_i	m_i	σ_i
C_0	−0.554			C_0	−2.143		
$X_1{}^a$	1.907	0.953	0.171	$X_1{}^b$	5.016	1.110	0.240
$X_2{}^a$	6.025	−0.827	0.201	$X_2{}^b$	3.524	−0.972	0.194
X_3	−0.285	0.211	0.276	X_3	1.644	0.211	0.276
$X_4{}^a$	1.595	64.19	3.665	$X_4{}^b$	−2.459	59.09	7.153
X_5	−6.845	−0.570	0.167	X_5	−5.359	−0.570	0.167
X_6	0.739	1.113	0.379	X_6	−1.058	1.113	0.379
X_7	0.988	23.25	4.428	X_7	2.712	23.25	4.428
R	0.787			R	0.899		
RMS	2.066			RMS	3.745		

R, correlations coefficient between experimental and regressed values.
RMS, root mean square of regressed error.
[a]This is the parameter obtained by the fabric property along warp direction.
[b]This is the parameter obtained by the fabric property along weft direction.
[c]29 samples are selected so as to cover the range of the fabrics for higher grade suit. Worsted: 71%, woollen: 29%.

Table 6 — The size deviation in finishing suits from Ito's data base

Condition			Size deviation (mm)
Back-centre-seam length:			
$EM1 > 8$	and	$80 > RT1 > 70$	$+3$–$+7$
$EM1 > 7.5$	and	$RT1 < 70$	$+6$–$+10$
$EM1 > 10$	and	$65 > RT1 > 45$	$> +10$
$S2 > 1.5^a$			-10–-13
Shoulder width			
$10 > EM2 > 9$	and	$65 > RT2 > 55$	$+3$–$+5$
$12 > EM2 >$	and	$70 > RT2 > 60$	
$EM2 > 12$	and	$70 > RT2 > 60$	$+5$–$+10$
$EM2 > 12$	and	$RT2 > 70$	$+3$–$+5$
$EM2 > 16$	and	$RT2 > 70$	$+10$–$+15$
$EM2 > 12^b$	and	$55 > RT2 > 45$	$+10$–$+15$

[a]Press shrinkage by HESC-FT-103A method.
[b]Woollen fabrics.

Table 7 — Operation notice for the workers on production line

Fabric KESF data				Notice for sewing worker on production line
EM2	>	9 and	(1)	Warning against the size error.
RT2	;	55–65		Be careful in the handling of fabric on steam-press machine.
EEM2	>	12 and	(2)	Use weaker mechanical pressure of the press.
RT2	;	70–80	(3)	And use a york following the indication already prepared.

machine maker, in Tokyo. A new sewing machine which has the automatic overfeed function is now being tested to investigate the machine control by the mechanical properties of sewn fabric. The principle of the overfeed mechanism of this machine belongs to an open-loop control principle. Accordingly, it leads to an error in the result of the overfeed from the designed value. This error is sensitive to the mechanical properties of the fabric sewn and this machine has a function of adjusting the overfeed by the input signal supplied to it. The matter to be solved at present is how the fabric property information is given to the machine to obtain the exact possible overfeed result. A result of a basic experiment carried out until now is as follows. Two pieces of fabrics were sewn together in the warp direction. The ratio, (top feed)/(bottom feed) of the machine was fixed at 4.5/2.5. The length of the sewn line was 150 mm and the mean value of the resulted overfeed was $150 \times 0.034 = 5.06$ mm ($CV = 2.05$) for warp directional sewing and for 45°, sewing $150 \times 0.117 = 17.55$ mm ($CV = 6.47$).

The equation predicting the overfeed has been derived in the same form as the equation (1) for the warp directional sewing, and the 45° biased sewing respectively. The equation is shown in Table 8.

The fabric mechanical parameters used in these equations were selected after eliminating those which are not significantly correlated with Y_0 or Y_{45} values. In conclusion, it has been obtained that the result is applied largely when a fabric has larger bending stiffness, less shear stiffness, and larger hysteresis component in bending. This result is common for both sewing directions, however, the effects of the shear stiffness and bending in elasticity on the result are more remarkable in the case of 45° biased direction than in the warp direction. If the machine is adjusted according to this prediction for individual fabrics, the scattering zone of the deviation will be reduced as predicted in Fig. 4, the prediction was made assuming the machine adjustment. The investigation is continuing to get the more narrow scattering from both sides, fabric and machine design. The solid line in Fig. 4 shows the distribution of the overfeed result obtained from 28 samples by using the Juki sewing machine in the 45° biased sewing experiment. The definition of the overfeed is length of the overfeeding fabric over the seamed state length of the seam line. The target overfeed is 18 mm and the seam length is 150 mm. The broken line is the distribution of the overfeed estimated by assuming that the control mechanism of overfeed of this machine corrects the predicted overfeed to the target value, before sewing, by means of fabric data input to the machine based on the equation for the 45° biased direction sewing as shown in Table 8. The distribution becomes narrow, however, still has considerable deviation for the target. Investigation for the improvement in the machine mechanism and the prediction accuracy is continuing.

3.3 The prevention of seam pucker

Even though the automatic overfeed machine performs exactly the designed overfeed, the seam puckering problem cannot be eliminated and still it remains. This puckering problem cannot be solved by the setting of the sewing machine only and it depends on the fabric mechanical properties. For the automated sewing system those fabrics for which danger of seam puckering is predicted, they must be excluded prior to their processing. Seam puckering occurs for three reasons as follows; (1) inappropriate mechanical property of fabric, (2) the relaxation shrinkage of fabric, (3) the poor setting of sewing machine. The relaxation shrinkage of the fabrics processed in the men's suit manufacturing factories in Osaka area is controlled to be less than 1.5% or, in some factories, 1%, and there is no more serious problem coming from fabric shrinkage at present. The first reason, however, is still serious and it is the main reason of puckering.

Our experiment which was carried out recently shows a clear correlation between seam puckering and the mechanical property of fabric. Seam puckering

Table 8 — Prediction equation[a] of resulted overfeed(unit: mm)

Warp direction sewing; Y_0				45° biased sewing; Y^{45}			
X_i	C_i	m_i	σ_i	X_i	C_i	m_i	σ_i
C_0	5.061			C_0	18.09		
$X_1{}^a$	−0.706	0.967	0.224	$X_1{}^b$	−1.116	1.027	0.310
$X_2{}^a$	−2.094	−0.889	0.275	$X_2{}^b$	4.156	−0.900	0.247
X_3	−1.563	0.195	0.207	X_3	4.099	0.180	0.196
$X_4{}^a$	1.247	65.71	8.232	$X_4{}^b$	−0.868	64.41	8.355
X_5	0.066	−0.545	0.248	X_5	6.091	−0.547	0.252
X_6	0.344	1.057	0.166	X_6	2.634	1.041	0.147
X_7	0.963	18.22	4.605	X_7	−1.803	20.21	5.687
R	0.728			R	0.823		
RMS	1.404			RMS	3.376		

[a]This is the parameter obtained by the fabric property along warp direction.
[b]This is the parameter obtained by the fabric property along weft direction.
[c]28 samples are selected so as to cover the range of the fabric weight for men's suit, W=31.04–12.07 mg/cm^2.

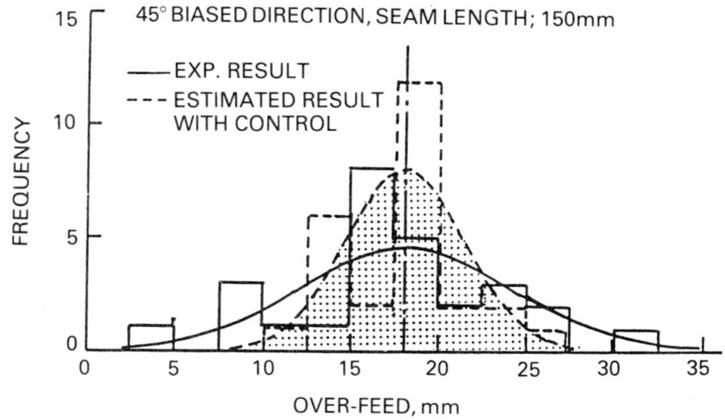

Fig. 4 — Solid line is the distribution of the overfeed error from the target value of 18 mm/150 mm. The broken line is the estimated distribution of the overfeed error when the machine is controlled based on fabric mechanical parameters.

caused by the fabric mechanical property appears most remarkably in a seam line when overfeed is applied. Although puckering can be prevented by machine setting to some extent, puckering caused by fabric mechanical property can not be controlled by setting of the sewing machine. And also, the difficult fabrics in puckering tend to increase in severity than normal fabrics when the machine setting is not perfect, that is, the fabrics having such difficulty in their mechanical property are sensitive to the machine setting.

Seam puckering associated with overfeed sewing appears remarkably at the seam line of shoulder arm hole, front-side seam and back-side seam lines. The model sewing as shown in Fig. 5 were carried out using various fabrics to seek the correlation between the degree of puckering and the mechanical property of these fabrics. Fabrics of 24 kinds were selected and 9 samples for each kind of fabric, that is, 216 samples in total were tested. Seam puckering along the sewing line was graded by the puckering value as shown in Table 9 in subjective evaluation. An example of the seam pucker of $P = 3$, which appears along the seam line of the curved sewing with overfeed, is shown in Fig. 6.

The sewn samples were pressed with a steam-press machine then kept for 24 hours in room condition. Then experts in the two factories of men's suit manufacturing evaluated the degree of puckering of these samples subjectively following the guide of Table 9. The first experiment was sewing along the warp direction without over-feeding. In this straight line of sewing, the two pieces of fabrics must result in a flat surface after sewing. The amount of seam puckering in this case is not so large but easily visible because of the flat surface. Using similar procedure of regression analysis as in case of the size change and the overfeed prediction, an equation predicting puckering value, P, has been derived and shown in Table 10. This equation indicates that properties such as less extensible and elastic in extension (higher RT), inelastic in bending and light fabric-weight are the main factors relating to puckering.

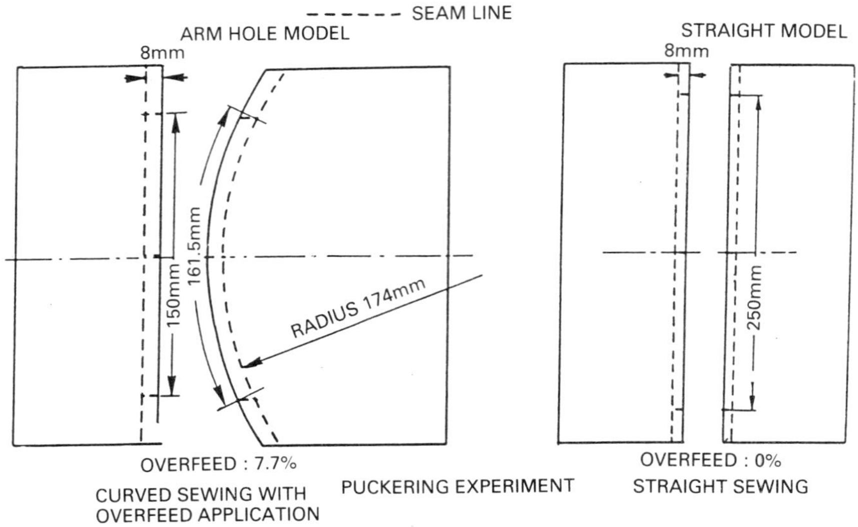

Fig. 5 — Seam pucker experiment using two models. Left: curved sewing with overfeed of 7.7%. This is a simulation of top arm hole part of suit. Right: straight line sewing without overfeed.

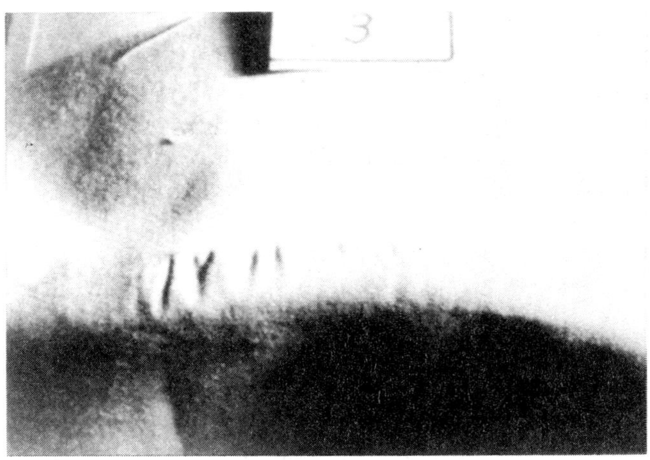

Fig. 6 — An example of the seam pucker of $P = 3$, which appears on the curved sewing with overfeed.

The next experiments for the puckering problem have been carried out using the two pieces of fabric having a straight and a circular line, respectively and these two lines were put together and sewn applying overfeeding to make a circular line

Table 9 — Puckering value (*P*-value)

P-value	Degree of puckering
0	not puckered
1	slightly puckered
2	puckered
3	terribly puckered

simulating the arm hole seam line in the shoulder top-part as shown in Fig. 5 and Fig. 6. The overfeed of 7.7% was applied exactly and uniformly along the seam line by manual operation regardless of the overfeed capacity of the fabric samples. The puckering value is correlated with the mechanical properties of the fabric samples and a regression equation has been derived as shown in Table 10. The correlation coefficient (R) between the regressed and the experimental values of P is 0.83.

From the two regression analysis in Table 10, puckering occurs easily in fabrics having properties such as, inextensible and highly elastic (high RT), stiff and inelastic in bending and shearing properties and light fabric-weight. The P value must be 0 or less than 1 for suits as commercial goods and if P is larger than 2, tailoring of the fabric must be considered to stop regardless of the performance of the sewing machine with no condition.

4. A CENTRE FOR FABRIC SPONGING USING KESF AND SHRINKAGE TESTING

The major 12 companies of men's suit manufacturing in Japan have their factories in the Hirakata area of Osaka and produce about 1.4 million men's suits of high class quality per a year. These companies have organized an association named Osaka Apparel Technology Modernization Association (OATMA). This association operates a centre for testing and sponging of the fabrics which are processed in these factories. The diagram in Fig. 7 shows the function of this centre. The fabrics bought by every factory in this area are sent to the centre at first for sponging. These fabrics are also tested for their tensile, shearing and bending properties by using KESF and also the steam-press–shrinkage test based on the HESC-FT-103A simplified method. Then the fabrics go through the four sponging lines as shown in Fig. 8. The selection of the sponging line is made according to the fabric properties tested prior to sponging. The sponging line B, which is used for most fabrics, about 80% of the fabrics processed in this centre, can control the degree of sponging by the amount of steam applied onto the fabric during sponging. The degree of sponging, that is, the amount of the steam is decided for every roll of fabric on the basis of the fabric mechanical and shrinkage data, and the data of the fabric is input to the sponging machine to give the optimum condition of sponging for every roll.

Table 10 — Prediction equation[c] of puckering value P

Arm hole model				Straight model			
X_i	C_i	m_i	σ_i	X_i	C_i	m_i	σ_i
C_0	0.729			C_0	1.440		
$X_1{}^b$	−0.131	1.088	0.310	$X_1{}^a$	−0.215	0.920	0.241
$X_2{}^b$	0.817	−1.124	0.298	$X_2{}^a$	−0.127	−1.029	0.224
X_3	0.426	0.142	0.186	X_3	−0.064	0.153	0.190
$X_4{}^b$	0.217	66.79	4.425	$X_4{}^a$	0.192	67.48	5.633
X_5	−0.978	−0.670	0.218	X_5	−0.384	−0.668	0.214
X_6	−0.293	0.104	0.133	$X_6{}^a$	0.154	1.025	0.158
X_7	−0.138	17.55	0.133	X_7	−0.020	17.63	4.492
R	0.834			R	0.840		
RMS	0.469			RMS	0.389		

[a] This is the parameter obtained by the fabric property along warp direction.
[b] This is the parameter obtained by the fabric property along weft direction.
[c] 24 samples are selected so as to cover the range of the fabric weight for men's suit, W=33.41–9.92 mg/cm^2.

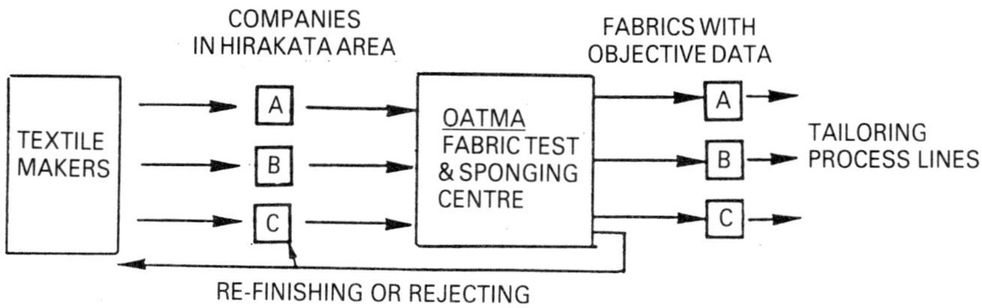

Fig. 7 — The centralization of objective measurement and sponging.

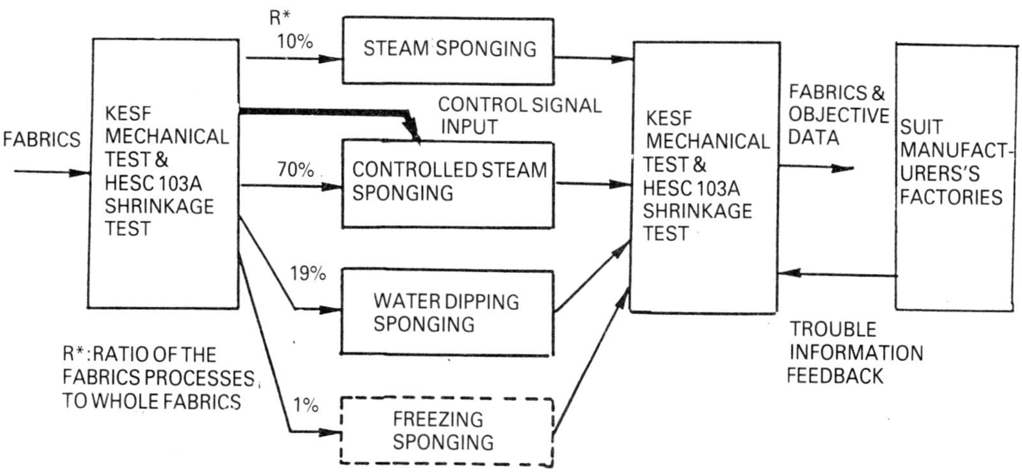

Fig. 8 — The central system of *OATMA* objective measurement and sponging.

This system is now being fully automated and the fabric data from the data processor of KESF is sent to the sponging machines directly to control it automatically. After sponging, KESF and steam-press testing are again carried out for all the finished fabrics, then the fabrics are sent to each client's factory with the KESF and the shrinkage data for their tailoring process control. This centre now covers all the factories in the Hirakata area and is now also increasingly covering other factories in the other parts of Japan. In the centre, two operators are working for both KESF and shrinkage tests and measure these properties for 40 samples per day on average.

Fig. 9 shows the suit (medium to higher grade) manufacturing map of Japan. Now, about 66% of all Japanese medium to high grade suits are under control based on the mechanical and shrinkage inspections. The rest 34% are produced by companies located in the Tokyo area and these companies are now starting individual testing systems using the same testing system as the Hirakata's system.

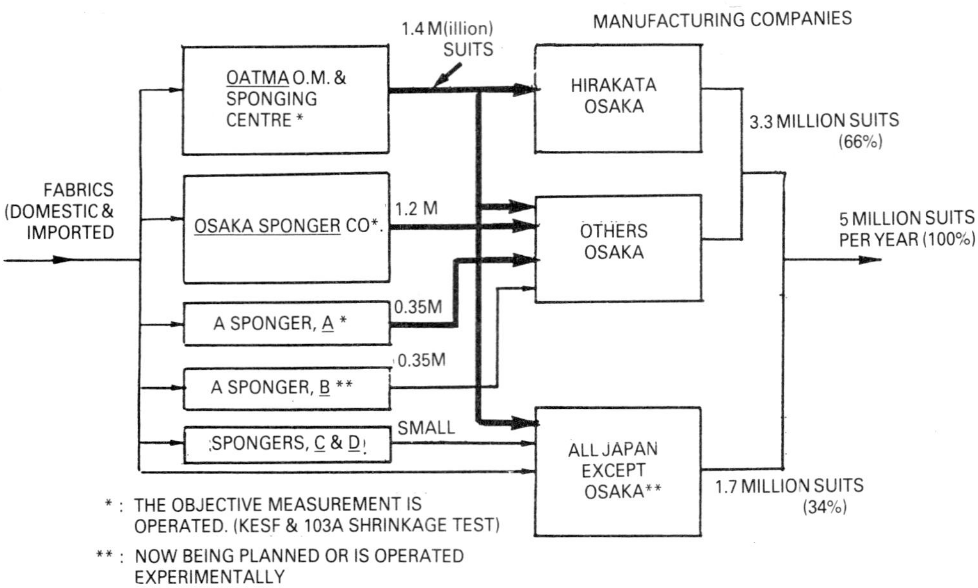

Fig. 9 — Map of the Japanese medium–higher grade men's suit production and the application of objective measurement at present — 1989. Thick-solid lines are the flow of the fabrics which come through the objective measurement inspection station.

The very high grade suits are made by factories in the Hirakata area and 100% of them use this control.

5. THE SPREAD OF THE OBJECTIVE MEASUREMENT FROM LABORATORY TO PRODUCTION SITE

About 15 years have passed since several factories of men's suit manufacturers in Osaka area started the use of the KESF data around 1975. Since then, a factory of Sanko Iryo Co., for example, already measured 120 thousand fabrics for their mechanical and shrinkage properties in these ten years. Some of the other factories of major companies have also similar numbers of testing. The method of the tailoring process control in every factory is different, not in principle but in the detail of actual procedures. This control system is based on its own database accumulated by itself for these 10 to 15 years. With increasing of the database, more detailed control is now becoming possible. On the other hand, some of KESF parameters are now being spotlighted in apparel engineers as follows.

At the beginning of the application of the KESF in the apparel field, fabric tensile property was highlighted by the apparel engineers who applied it, because this property is considered to be closely related with difficulty in the tailoring processing, but the fabrics produced in many textile mills have been improved gradually in this property according to the feedback, from the apparel engineers with close communication between the textile and the apparel factories using the common objective data and language concerning fabric properties. The parameters being highlighted recently are the parameters concerned with the hysteresis component in the mechanical properties. For example, $2HG5$ which is the hysteresis in shear force at 5 degrees of shear deformation angle is becoming the parameter which many apparel engineers commonly used for checking the fabric property for their tailoring process and also for preventing consumers' claims about their products as follows.

When

$2HG5 > 3.5$	Wrinkle problem will occur. The consumer's claim to wrinkle of suits will increase rapidly if such a fabric is tailored.
3.0–2.5	Probably the upper limit of the acceptable fabrics for apparel factory.
0.5–1.3	The danger range to the seam puckering of suits, but the suit will be an excellent suit if puckering is prevented by appropriate control.
$2HG5/G > 1.8$	The appearance of the suit from these fabrics will become poor.

On the other hand, higher $2HG5$ is better for the lining fabric. The lining fabric makers are improving their lining material products based on the guide lines supplied from the apparel factories. Such sorts of common experience recognized by apparel engineers and are communicated in the apparel industries to assist them as a quick evaluation method of fabric performance just like in the handle judgement.

According to the shifting of fabric objective measurement from laboratory to production site, the KESF instrument, which was designed for laboratory analysis originally, is under the necessity of being modified. The most popular component of the KESF series for apparel industries is the tensile/shear tester called KESF-B1. A new automated KESF-B1 AUTO is now being completed for use at production sites. The operation for measurement using this equipment is fully automated, the operation is made by a push button. This automated operation can also eliminate the error in measurement caused by the skills of operator completely because of the automated operation.

6. THE NEAR FUTURE OF APPAREL ENGINEERING

The use of fabric objective measurement in the apparel industry has been rapidly spread in Japan. There are some other reasons which have not been discussed here. The one is the increasing variation of fabric type for suits such as the blend of three or more fibre components of synthetic and or natural fibres, very light weight suiting with fine wool, use of surface treatment chemicals such as silicon softener, etc. The tailor's experience based on traditional fabrics is no more applicable to their complexity. Another reason is that, as mentioned before, the suit tailoring has changed its structure from the craftsman's tailor shop style to line production system and the fabric market has become more international. Many fabrics are imported from many textile companies in different countries. It is hard for apparel engineers to catch the character of every fabric quickly. In the past, tailors have bought fabrics through small local markets and tailors knew the fabric characteristics very well. This situation will continue in the future and the fabric objective measurement will become more necessary.

Automated sewing machines will increase their use gradually in the apparel industry to improve the labour-intensive style of apparel manufacturing. The intelligent sewing machines which can be controlled by fabric properties will become more popular in the near future. The automation of individual machines in the production line is much more important than the completely continuous automated line at this stage. First the completion of the component machine is necessary and then a total system can be designed.

The measuring system for fabric mechanical properties, which have been developed originally for objective fabric handle evaluation, have been directly applied in the apparel industries beside the fabric handle evaluation and its technology has become very popular in the last 10 years, however, the objective handle evaluation technology has not been sufficiently applied in the apparel industries. In the future, however, probably very near future, the application of the handle evaluation technology will be applied in this industry to help them in the selection of the fabrics for the higher quality suit. This will be the second stage of the application of the fabric objective measurement technology in the apparel industry.

ACKNOWLEDGEMENTS

We would like to thank, Sanko Iryo Co., Oga Clothing Hirakata Co. and Juki Co. for their cooperation.

REFERENCES

[1] S. Kawabata; 'The Standardization and Analysis of Hand Evaluation', HESC, the Textile Machinery Soc. Japan, Osaka, Japan, 1975, 2nd edition, 1980.

[2] S. Kawabata; in *Proceedings of 1st Japan/Australia Symposium on Objective Specification of Fabric Quality, Mechanical Properties and Performance* (edited by S. Kawabata, R. Postle and M. Niwa), the Textile Machinery Soc. Japan, Osaka, Japan, 1982, p. 31.

[3] M. Inoue; in same literature as [2], p. 317.

[4] HESC; 'HESC-FT-103A, The Testing Method of Dimensional Instability of Fabrics', *J. Textile Machinery Soc. Japan*, 1979, **32**, p 284.

[5] S. Kawabata; 'A Report on the Development of the Testing Method. HESC-FT-103A for Dimensional Stability of Clothing Fabrics Caused by Steam Pressing', *J. Textile Machinery Soc. Japan*, 1979, **32**, p. 268.

[6] K. Ito; in *Proceedings of 2nd Australia/Japan Symposium on Objective Evaluation of Apparel Fabrics*, (edited by R. Postle, S. Kawabata and M. Niwa), the Textile Machinery Soc. Japan, Osaka, Japan, 1983, p. 89.

[7] M. Nitta; in same literature as [2], p. 321.

[8] M. Niwa, S. Kawabata and K. Ishizuka; in same literature as [6], p. 67.
[9] M. Niwa and S. Kawabata; in *Application of Mathematics and Physics in the Wool Industry*, WRONZ, 1988, p. 404.
[10] S. Kawabata and M. Niwa; *J. Text. Inst.,* **80**, 1989, p. 19.
[11] K. Ito and S. Kawabata; in *Proceedings of 3rd Japan/Australia Symposium on Objective Measurement: Applications to Product Design and Process Control* (edited by S. Kawabata, R. Postle and M. Niwa), the Textile Machinery Soc. Japan, Osaka, Japan, 1985, p. 175.
[12] M. Uemura; in same literature as [6], p. 387.

APPENDIX: KESF FABRIC MECHANICAL AND SURFACE PARAMETERS MEASURED UNDER THE STANDARDIZED CONDITIONS [1,2]

Note 1: EM was not included in the original sixteen parameters that were used for the prediction of hand values, but it has recently often been used to predict fabric-formability performance in tailoring.

Note 2: Tensile, bending, shearing, and surface properties have both warp and weft directional values for each of their parameters. Suffix 1 and 2 indicate the warp and weft directional values, respectively, such as LT_1, WT_1 and so on. The absence of a suffix indicates the average value of the two directional values.

Parameters		Description	Unit
Tensile	LT	Linearity of load/extension curve	none
	WT	Tensile energy	gfcm/cm^2
	RT	Tensile resilience	%
	EM	Extensibility, strain at 500 gf/cm of tensile load	none
Bending	B	Bending rigidity	gfcm^2/cm
	$2HB$	Hysteresis of bending moment	gfcm/cm
Shearing	G	Shear stiffness	gf/cm.deg
	$2HG$	Hysteresis of shear force at 0.5° shear angle	gf/cm
	$2HG5$	Hysteresis of shear force at 5° shear angle	gf/cm
Compression	LC	Linearity of compression/thickness curve	none
	WC	Compressional energy	gfcm/cm^2
	RC	Compressional resilience	%
Surface	MIU	Coefficient of friction	none
	MMD	Mean deviation of MIU	none
	SMD	Geometrical roughness	μm
Thickness	T	Fabric thickness	mm
Weight	W	Fabric weight	mg/cm^2

Details of the deformation modes and measuring conditions for fabric mechanical, surface, and constructional properties

Measuring conditions

Properties	Measuring conditions	Parameters
Tensile	Strip biaxial deformation. Maximum load: F_m=500 gf/cm. Tensile-strain rate: 0.4%/s (speed: 0.2 mm/s) Sample width: 20 cm. Sample length in the extension direction: 5 cm	LT_1, LT_2 WT_1, WT_2 RT_1, RT_2 EM_1, EM_2
Bending	Pure bending. Maximum curvature: $K_{max}=\pm 2.5$ cm^{-1}. Bending-curvature rate: 0.5 cm^{-1}/s	B_1, B_2 $2HB_1$, $2HB_2$
Shearing	Shear deformation under constant tension: W=10 gf/cm. Maximum shear angle: $\phi_m=\pm 8°$. Rate of shear strain: 0.00834/s (speed: 1°/s). Sample length: 5 cm. Sample width: 20 cm. Shear deformation is applied to the width direction	G_1, G_2 $2HG_1$, $2HG_2$ $2HG5_1$, $2HG5_2$
Compression	Compression in fabric thickness by two circular plates having area 2 cm^2. Maximum pressure: P_m=50 gf/cm^2. Rate of compression: 20 μm/s	LC WC RC
Surface	Contactor for friction measurement: ten parallel steel piano wires with 0.5 mm diameter and 5 mm length simulating finger-skin geometry. Contact force: 50 gf	MIU_1, MIU_2 MMD_1, MMD_2
	Contactor for geometrical roughness: a steel-piano-wire with 0.5 mm diameter and 5 mm length, placed on fabric parallel to fabric surface with contact force of 10 gf. Measured distance: 2 cm. Velocity: 0.1 cm/s. Tension on specimen: 20 gf/cm. The signal of wavelength less than 1 mm is allowed to pass by a high-pass filter of 2nd order	SMD_1, SMD_2
Thickness	Thickness at 0.5 gf/cm^2 pressure	T
Weight	Mass of specimen per unit area	W

QUESTIONS

STYLIOS: Seam pucker is caused by the interaction of the sewing process, the sewing thread and the fabric, and it seems to me that when we talk about objective measurement as far as fabric is concerned under this particular problem we should take into account the objective measurement of the thread as well and also the dynamics of the sewing machine. Because we all know that if we have a drop feed sewing machine for example which is spring loaded, at some particular speed the foot will bounce, inducing the problem.

KAWABATA: You are right. Seam pucker relates to fabric properties and sewing machine conditions. So you have to consider these two factors. In this research I have used one machine; of course you have the variation of one machine to another, one equation accounts for one machine; if you use a different sewing machine you have to use another equation to account for it. However, the trend of the equation that I have shown, I think, is similar for any sewing machine. This equation is flexible and can change as you pointed out.

JAYARAMAN: Do you have a database of all the collected information?

KAWABATA: Yes and it is a very big one. The centre that I have talked about processes about 1 million suits, it is a big number. All data is stored in a computer, so that they can compare the data.

JAYARAMAN: Are you extending the database by inputting the solutions as well or are you only storing the problems?

KAWABATA: We store these problems and separate the textile data from the garment data. This is particularly important for textile producers who do not know the problems of the apparel process.

SARHADI: You have shown some equations which you get from your raw data to the handle value. Are these empirical or theoretical equations and are these standard equations? How are they derived from? Or if someone comes up with a different coefficient will he get a different result?

KAWABATA: The equation is derived from the expert judgement. The meaning of standardization is difficult. We use one main equation very frequently and very often, so this tells you something about its accuracy. The main equation is for men's winter suits, there is another equation for men's summer suits. These two are the main equations but many for predicting quality, i.e. others for jackets others for pants, tracks, etc.

SARHADI: So one would have a vast array of equations for various applications and many varieties on top of that.

KAWABATA: The most important equation is for men's winter suits. Recently we have found applications of the equation for apparel fabrics in artificial leather, paper and so on, because in all cases you touch the material by hand. This is a common way to evaluate the properties, so we can apply the men's suit equation. In this case we already use the normalized value for each mechanical parameter, so we have to use the standard deviation and mean values in the equation for the artificial leather values and paper values. We can change easily this equation to fit the application.

ASSOUSA: I would like to express my personal thanks and to tell you how impressed we are for you working on the leading edge for our industry — you, Professor Niwa and your colleagues. Having said that, I would like to ask you if you could shed some more light now on the commercial application of your KES Auto system to the industry today or are you in the experimental stage?

KAWABATA: The principles behind the development of the KES Auto are well known. There are companies that need to test a lot of samples per day, so they need something faster. Also the variation of inserting the sample in the instrument jaws, between person to person is reduced dramatically, when using the automatic system. Therefore, it is more appropriate for some centres to use the new system, than others.

Introduction to the next paper

The work reported in the next paper reflects upon the work of the team in the Philadelphia College of Textiles and Science which is to continue until Autumn and hopefully extended. Professor Barndt did the experimental part of this work, Professor Weiner the sewing trials and the collaboration with industry, and Professor Fortess developed the project. Unfortunately Professor F. Fortess cannot be here with us to deliver this paper. I would deliver it instead but I only know the principle about this work and more about the financial arrangement of it. The title of the paper is 'The use of KES and FAST instruments in predicting processability of fabrics in sewing'.

The use of KES and FAST instruments in predicting processability of fabrics in sewing

Herbert Barndt
Director, Grundy Textile Product Evaluation Laboratories, Philadelphia College of Textiles and Science, USA
Fred Fortess
Director, Apparel Research Services, Philadelphia College of Textiles and Science, USA
Mel Wiener
Director, Seam Engineering and Sewability Laboratory, Philadelphia College of Textiles and Science, USA
J. Cyril Furniss†
Dean, School of Textiles, Philadelphia College of Textiles and Science, USA

ABSTRACT

This paper describes the experience of the Philadelphia College of Textiles and Science with the use of the Kawabata Evaluation System for Fabrics (KES-F) since 1984 and more recently, 1988, with the Fabric Assurance by Simple Testing (FAST). Although previous projects have involved the effect of *chemical and mechanical finishing* and *sponging* on the *hand* and *mechancial properties of fabrics*, the major emphasis in this paper correlates the *processability of fabrics* in tailored clothing manufacturing, especially in *manually operated overfeed sewing*, with the *tensile elongation* and *shear properties* of these fabrics. It is anticipated that *programmable sewing machines* will require instructions based on the mechanical properties of the fabrics being processed.

INTRODUCTION

Any conference concerned with the objective measurement of the mechanical engineering properties of textile structures must recognize its indebtedness to the pioneering work of F. T. Peirce [1] published in the *Journal of The Textile Institute*, vol. 21, pp. T377–416, 1930.

† Who presented the paper.

In the paper 'The "Handle" of Cloth as Measurable Quantity', Peirce translated the subjective physiological sensations of stiffness, hardness and smoothness into such mechancial properties as bending length and flexural rigidity. Subsequently, J. Lindberg, L. Waesterberg, and R. Svenson [2] in their paper 'Wool Fabrics as Garment Construction Materials', first introduced the concept of fabric formability as a measure of the degree of compression in the plane of the fabric before buckling occurs.

Although individual researchers in Europe, Australia and Japan, mostly at universities and in Government research centres, continued to explore the measurement of the mechanical properties of fabrics, the major involvement of textile and apparel manufacturers came as a result of Japanese Government legislation encouraging modernization of small businesses. As early as 1961, the men's suit makers formed the 'Osaka Modernization Cooperation of Men's Suiting', with the objective of improving the technology and management of small clothing manufacturers. By 1972, through the effects of Dr Sueo Kawabata and his associates, and the support of the Textile Machinery Society of Japan, it was possible to create the Hand Evaluation Standardization Committee (HESC). H. Inoue [3], Melbo Clothing Company, Ltd., in his paper in the first 'Joint Symposium on the Objective Specifications of Fabric Quality, Mechanical Properties and Performance', reviewed the Japanese industry's shift in interest from the shrinkage properties and the hand of fabrics, to increased efforts to predict and control the processability of suiting fabrics, especially in sewing.

Attention must be given to the three symposia with its more than 100 papers, many of which reported on manufacturers' experience with the use of the KES-F:

Objective Specification of Fabric Quality Mechanical Properties and Performance (1982), Kyoto
Objective Evaluation of Apparel fabrics (1983) Victoria, Australia
Objective Measurement: Applications to Product Design and Process Control (1985) Kyoto, Japan.

These symposia are still available through the Textile Machinery Society of Japan, Osaka Science and Technology Center Building, 8–4 Utsubo Hommachi-1-Chrome, Mishi-Ku, Osaka 555, Japan.

In addition to the fundamental articles by Dr Sueo Kawabata and Dr Masako Niwa and their co-workers, and Dr Ronald Postle and his co-workers at the University of New Soth Wales and in the CSIRO laboratories, special attention should be given to the work of: M. Nitta, Oga Clothing Hirakata Company, M. Uemura, Melbo Clothing Company, and K. Ito, Sanko Iyro Clothing Company.

MISCELLANEOUS PROJECTS USING KES-F

Beginning early 1984, our objective at PCT&S was to explore the versatility of the use of the KES-F instruments. Table 1 lists several of the projects. Item 1 involved the effect of dyeing time on 100% wool fabrics of very weak shades (black and navy) and light shades (grey and tan). Clothing manufacturers had noted that frequently,

Table 1 — Miscellaneous projects using KES-FAST instruments

1. Effect of length of time in dyeing of light shades (tans and greys) and dark shades (navies and blacks) on some fabric construction in 100% wool.
2. Change in mechanical properties of fabrics and hand after sponging to reduce potential shrinkage.
3. Effect of application of durable silicone topical finishes on mechanical properties and hand of tailored clothing and denim fabrics.
4. Comparison of 'softness' properties of non-woven facial tissue structures.
5. Evaluation of fabric softness on 100% cotton terry towel fabric.
7. Sewability of coating fabrics on the Singer-Draper Robotic Back Seam and Vent Machine.

very dark shades processed differently than very light shades. Significant differences were observed in the mechanical properties of these fabrics which were not easily corrected by finishing.

Item 2 involved the change in mechanical properties of fabrics in sponging of wool containing fabrics. Clothing manufacturers were particularly concerned with the change in the subjective perception of the hand of the fabrics after sponging, regardless of the sponging procedure used. However, it was noted that in some cases, the hand was improved and in some cases where the hand was less desirable, the manufacturer would attempt to improve the hand by adding a finish.

Items 3, 4 and 5 involved the application of chemical finishes to improve the perception of softness. Of special interest was the correlation of the KES-F data with hand panel perception of the softness of facial tissue and the softness and fluffiness of terry towels.

Of particular interest was item 6 which requested that we predict which fabrics would process satisfactorily on the Singer-Draper Robotic Coat Back Seam and Vent Machine.

No information was provided with respect to the fabric construction or fibre content of the five fabrics which were submitted. In Table 2 are recorded the tensile-elongation data EM1, EM2, RT and LT and only on fabric, #5 navy blue, provided data which would be considered as meeting the non-control zone in H. Ito's Table 3, which was based on observations made on thousands of suiting fabrics over a 10-year period. This was the only fabric of the five which produced acceptable seams. It is important to note that all five fabrics were commercially used and produced acceptable seams on manually processed sewing machines where the operator could manipulate the fabric and/or adjust the sewing machine to be compatible with the differences in the fabrics being processed.

Table 2

Fabric	EM_1 EM_2 % Elongation		RT % Resiliency		LT Linearity	
	W	F	W	F	W	F
1. Navy blue	3.9	2.3	46.0	52.5	0.68	0.87
2. Light silk	2.2	6.3	54.0	56.6	0.67	0.51
3. Dark silk	2.5	5.0	55.6	52.1	0.64	0.54
4. Dark grey pin stripe	3.0	3.2	66.4	67.8	0.68	0.64
5. Blue navy shadow warp stripe	5.2	5.3	71.6	70.7	0.56	0.58

THE BEN FRANKLIN PROJECT: PREDICTING FABRIC PROCESSABILITY IN SEWING

Although the KES-F instruments have been applied in a great variety of fabric quality, product development and manufacturing processing evaluation projects, in this paper our interest is primarily concerned with predicting the processability of fabrics, especially in sewing. This interest was stimulated by the more than 10 years of use of the KES-F by the major Japanese tailored clothing companies which resulted in a number of papers included in the three Australia–Japan Symposia. In particular, the work of K. Ito, at the Sanko Iyro Clothing Company, resulted in the Tailoring Process Control Chart based on fabric tensile and shear properties recorded in Table 3.

Based on KES-F testing of thousands of fabrics, Ito was able to predict which fabrics would require special manipulation by the sewing machine operators or require adjustments in the sewing machine, in order to prevent seam puckering, problems in overfeed sewing and unacceptable appearance in the finished garment.

In our contact with the tailored clothing industry, through the Clothing Manufacturers Association-USA and the International Association of Clothing Designers, we have been aware of manufacturers' complaints about the inconsistency of fabric quality and the unexpected problems in manufacturing, especially in sewing and pressing. Since 1984, we have used the KES-F on many of these problems.

As a result of persistent requests for help, a project proposal was made to the Ben Franklin Partnership Fund of the State of Pennsylvania to investigate the measurement of the mechanical properties of fabrics in predicting fabric processability in sewing. As was our practice in our research on fusing and sponging, a consortium was formed which included three major clothing manufacturers, After Six, Greif and Brooks Brothers. By 1989, we had also acquired the Fabric Testing by Simple Testing (FAST) system developed by CSIRO of Australia and included its use in the Ben Franklin Project for comparison with KES-F. Table 4 is the FAST Control Chart used for comparison with KES-F Control Chart (Table 3).

Table 3 — tailoring process control based on fabric tensile and shear property

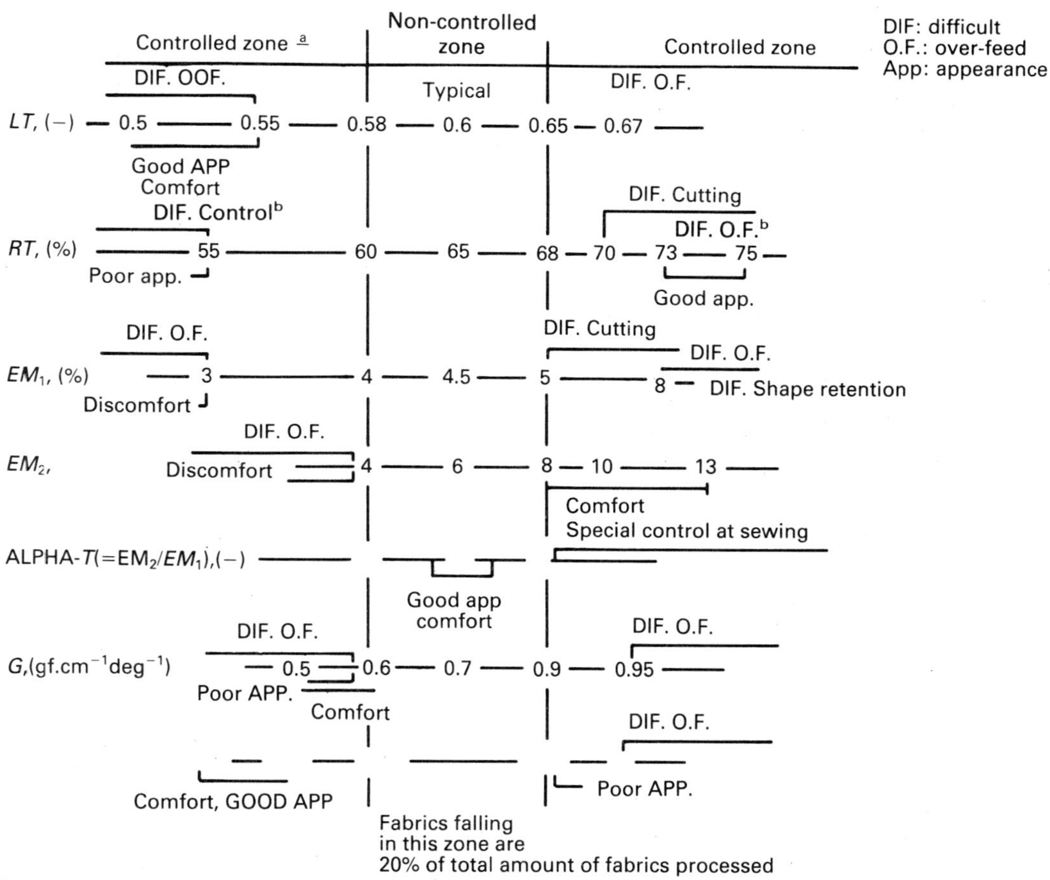

DIF: Difficult. O.F.: Over-feed. APP: Appearance.

[a] Controlled operation on the basis of fabric mechanical properties is required for conventional tailoring process.

[b] Especially difficult in the cases: $LT<0.55$ and $RT>73$ or $LT<0.55$ and $RT<55$.

The procedure in this project was to receive two-yard sponged samples from the consortium members for the mechanical properties reported on the control charts. The clothing manufacturers were responsible for observing and recording the processing behaviour of each of the fabrics in production.

EXPERIMENTAL PROCEDURE

Experimental work

A_2 series of wool and polyester–wool blending fabrics ranging from 150 g/m^2 to 250 g/m^2 were collected and evaluated using both the FAST and KES instruments. In the

Table 4 — The FAST control chart for light-weight suiting fabrics

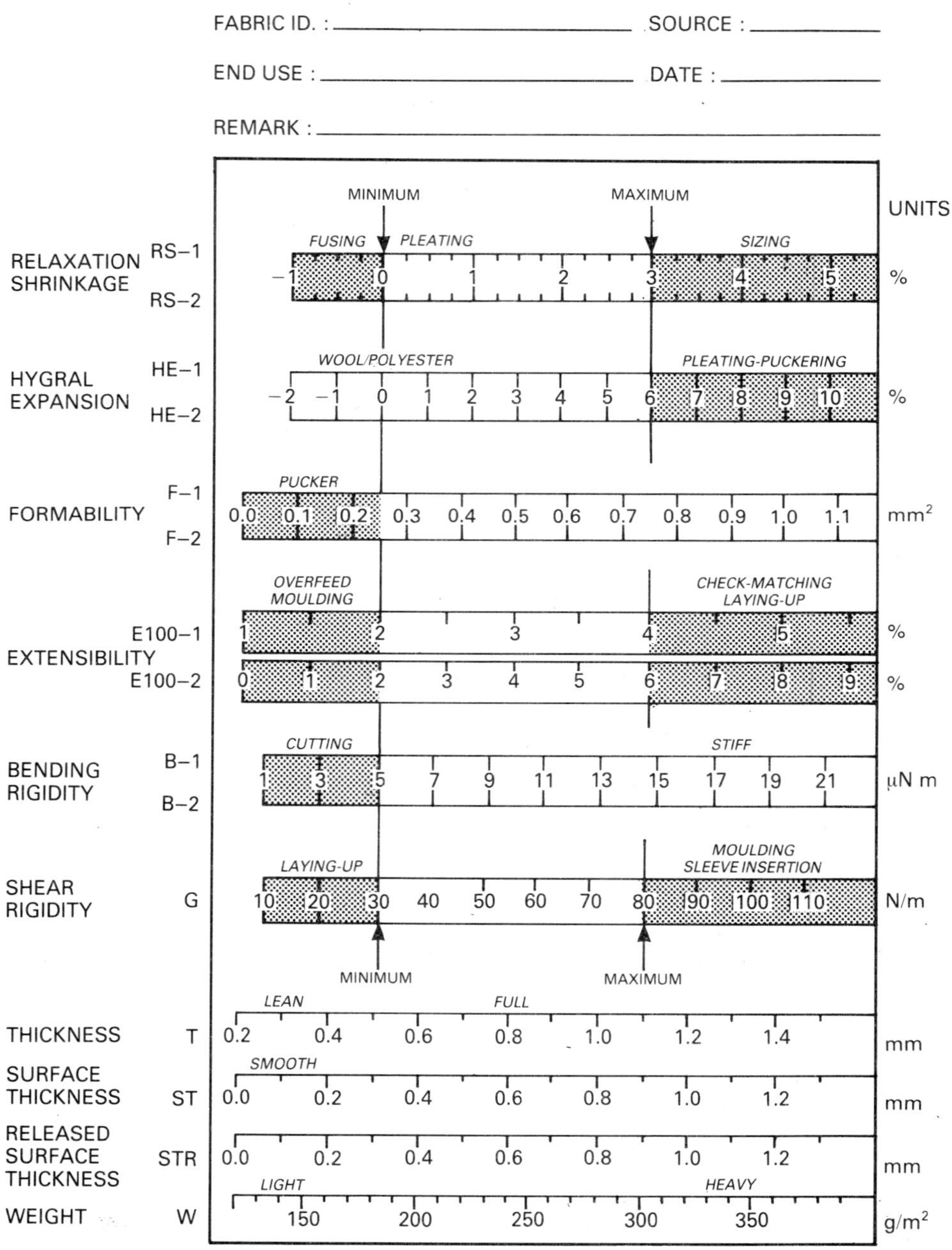

case of the KES instruments, procedures described by Ito were followed in establishing a fabric profile of acceptable mechanical properties for tailoring. These measurements were made on the KES FB-1 tensile and shear testing device. Details of these evaluations are described in Tables 5 and 6. Data was plotted to determine whether it

Table 5 — KES properties measured

LT (non)	— Linearity of the stress–strain curve.
RT (%)	— Tensile resilience measured by the load–extension curve.
EM (%)	— The maximum strain at a tensile load of 500 gf/cm.
Alpha (non)	— The ratio of EM_2 to EM_1.
G (gf/cm per degree)	— Shear stiffness.
2*HG*5 (gf/cm)	— Shear hysteresis at 5°.

Table 6 — FAST properties measured

RS (%)	—	Relaxation shrinkage.
HE (%)	—	Hygral expansion.
F (mm^2)	—	Formability.
*E*100 (%)	—	Extensibility at a load of 100 gf/cm.
B (μN.m)	—	Bending rigidity.
G (N/m)	—	hear rigidity.

fitted into the 'Non-Control Zone'. As established by Ito and Kawabata, the 'Non-Control Zone' describes fabrics with mechanical properties that will permit tailoring without special attention in sewing. Plots of four fabrics are presented in Fig. 1.

The same fabrics were evaluated using the FAST instruments. The FAST system consists of the following four devices: FAST-1 Compression Meter, FAST-2 bending Meter, FAST-3 Extension Meter, and FAST-4 Dimensional Stability Test. Details of these evaluations are described in Table 2. In addition, a 'Quality Control Chart' is provided and minimum and maximum values are suggested for properties evaluated. Plots of the same four fabrics referenced above are presented in FAST 'Quality Control Chart' format in Fig. 2.

Seams were prepared using the same fabrics. Details of the sewing conditions and seam orientation are described in Table 7. The appearance of the seams after sewing and pressing was evaluated and rated on a scale of 1 to 5, where a rating of 5 indicates the best appearance and a rating of 1 indicates poor appearance. The results of these ratings are included in Figs 1 and 2.

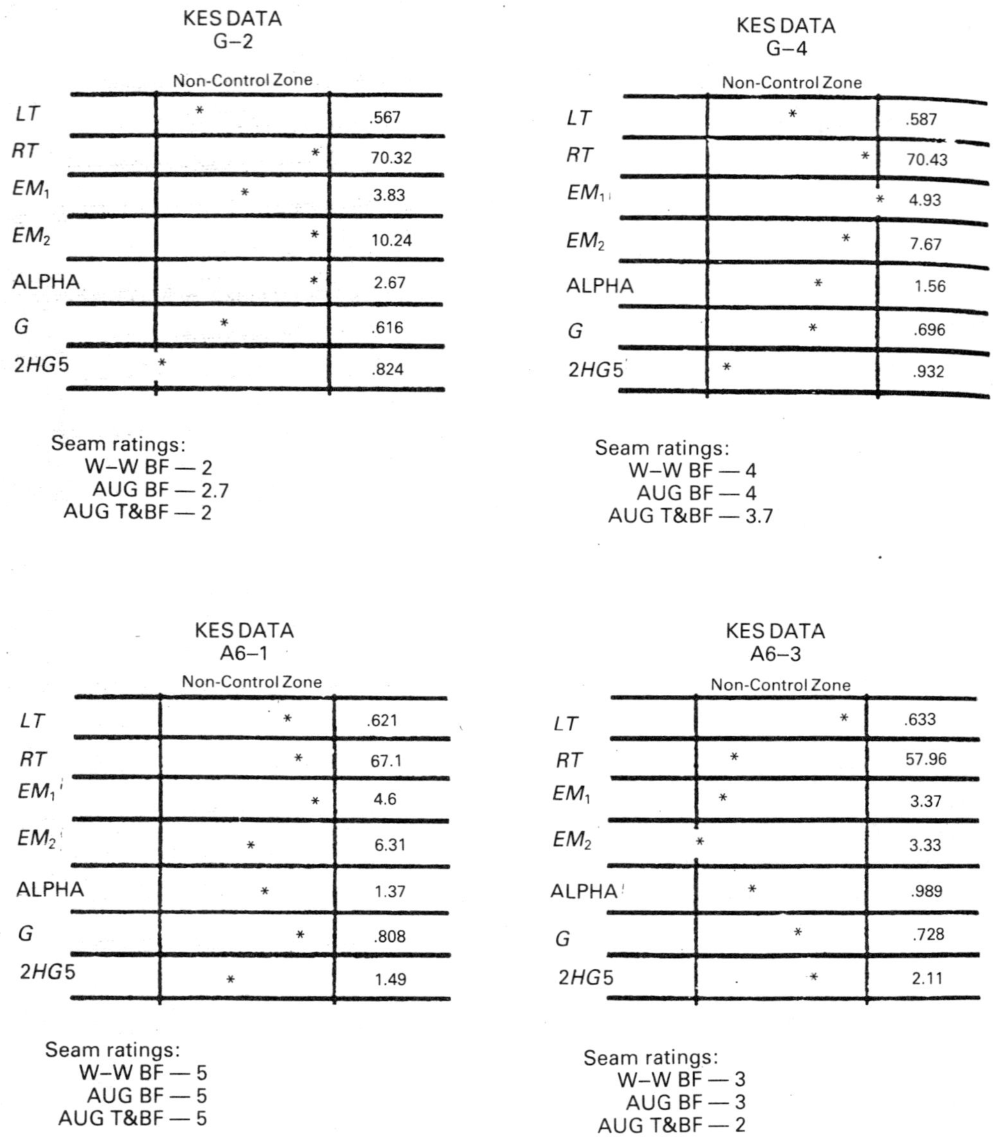

Fig. 1 — KES plots.

Discussion

Fabric identification G-2 fits the KES profile very nicely; however, the seam appearance is unacceptable. If we examine the FAST chart for the same fabric, we can see that values for relaxation shrinkage (RS-1 & RS-2) as well as filling hygral expansion (HE-2) are all well above recommended maximums. It is the fabric's instability (hygral expansion) in the presence of moisture that is causing the puckering at the seams. Fabric G-4 fits both the KES-FAST profiles and the seams

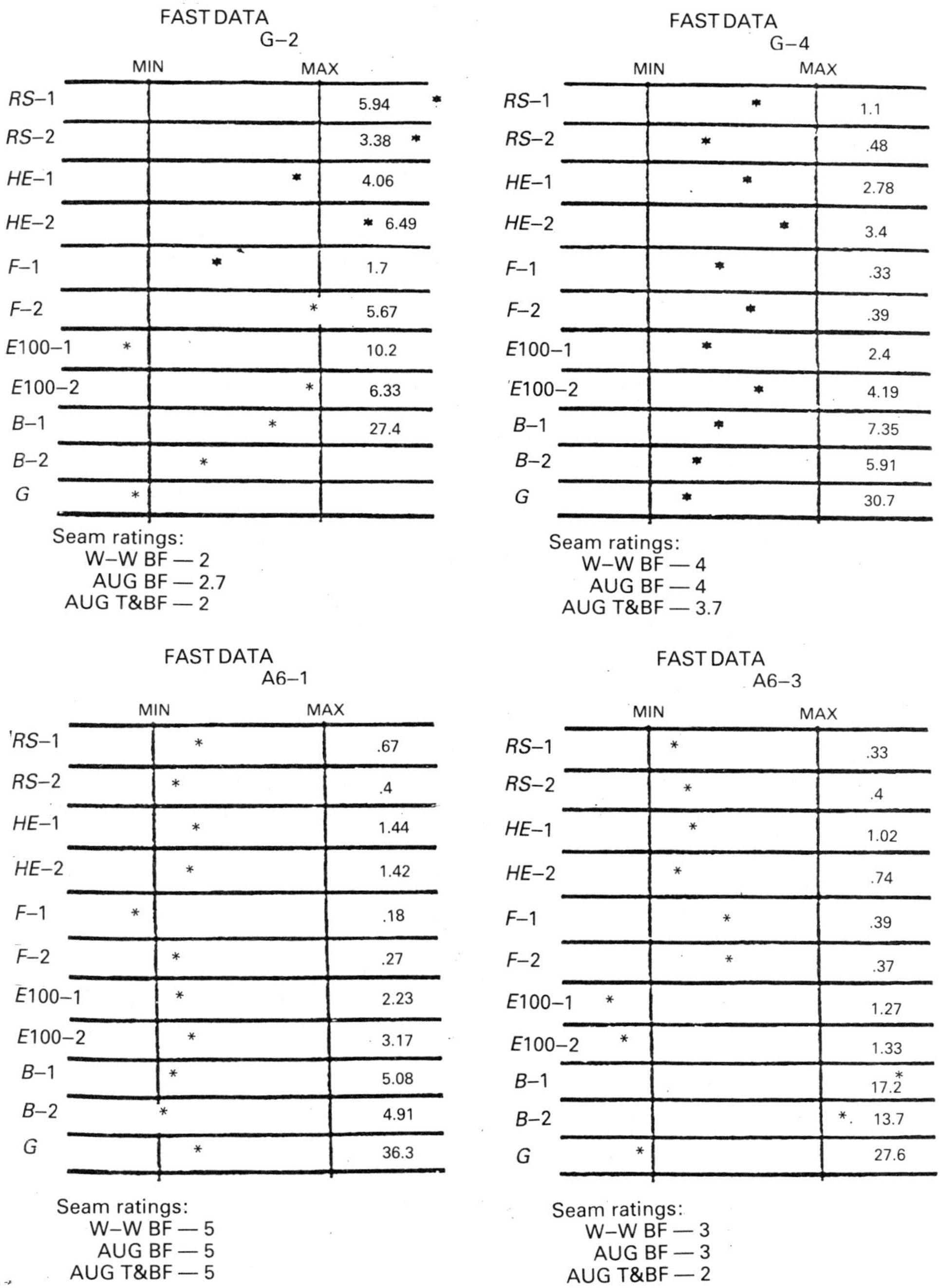

Fig. 2 — FAST plots.

Table 7 — Sewing conditions and seam configurations

Thread	—	30 tex, polyester core–cotton wrap.
Needle	—	DB × 1 Size 14.
Stitches	—	4.5 to 5.0 per cm.
Machine 1	—	Juki DDL — 5550-6 (bottom feed) BF.
Machine 2	—	Juki DLU — 5490-6 (top and bottom feed) T&BF.
Fabric arrangement		
W–W		Warp to warp
W–F		Warp to filling
F–F		Filling to filling

Note: Seams are described by arrangement and type of feed. Example: W–W BF 9 Warp to warp with bottom fed.

produced from this fabric had acceptable appearance. Fabric A6-1 produced the best seam appearance and it is clear that it fits both the KES and FAST profiles. The fact that warp formability (F-1) is slightly below the recommended minimum value, as depicted on the FAST chart, is not considered a serious problem. Fabric A6-3 produced marginal seams. While this fabric fit the KES profile, it can be seen that the FAST results for extension (E100-1 & E100-2) and bending (B-1 & B-2) are outside the recommended limits.

Conclusions

Based on a limited sample, fifty fabrics, it appears that the FAST system is a better predictor of tailoring difficulties than the system developed by Ito, using the KES. If fabric dimensional stability data is included with the measurements made on the KES, then it becomes a more sensitive indicator of fabric tailoring problems. Modification of the suggested limits established for either system can make the system more responsive to the needs of a particular manufacturer.

The Ben Franklin Project had as its objective to alert manufacturers to recognizing that 80% of their fabrics will require some special handing by the sewing machine operators. Experienced sewing machine operators usually know, without instructions from supervisors, how to handle difficult fabrics in order to achieve acceptable seams and garment appearance. Experience is already available which indicates that the mechanical properties of the fabrics being processed on automated machines will require that the machines be adjusted to be compatible with the fabric's properties.

REFERENCES

[1] F. T. Peirce, *Journal of Textile Institute*, 1930, **21**, T377–416.
[2] J. Lindberg, *et al.*, *Journal of Textile Institute*, 1960, **51**, 1475.
[3] H. Inoue, *Proceedings of the Joint Japan–Australia Symposium on the Objective Specifications of Fabric, Mechanical Properties and Performance* (1982).

[4] *Conception of the Automated Tailoring Controlled by Fabric Objective-Measurement Data*, K. Ito and S. Kawabata 1986, The Textile Machinery Society of Japan.
[5] *Fabric Assurance by Simple Testing*, CSIRO 1988, CSIRO Division of Wool, PO Box 7, Ryde, NSW, 2112, Australia.

Introduction to Professor Rosan Shishoo

It is my pleasure to introduce to you Professor Rosan Shishoo who was educated at Bombay University, going on to Sweden where he was awarded his Ph.D. at Charmers University of Technology. He started his professional career at HATRA in 1963; he joined TEFO in 1965 as a research scientist where his present position is director of research; this is coupled with him being Professor of Fibre Science and Technology at Charmers University of Technology in Gothenburg. He has published numerous papers on general topics on the physical and mechanical properties of fibres, yarns and fabrics. He is a fellow of the Textile Institute, a member of the Fibre Society of the USA, and a frequent speaker to many International Conferences. In 1988 he was awarded the Swedish Government's annual innovation prize for his contribution to science and technology. It gives me great pleasure to welcome Professor Shishoo who is going to talk about the 'Relation between fabric mechanical properties and garment design and tailorability'.

Relation between fabric mechanical properties and garment design and tailorability

R. L. Shishoo
Swedish Institute for Textile Research (TEFO), Göteborg, Sweden

ABSTRACT

Future *automation* and *robotization* of apparel manufacturing processes will undoubtedly require that the machines and systems are selected based on the specific properties of the fabric being processed. Thus it is essential to develop and use *objective evaluation methods* for producing *fabric compatibility data* which are necessary for control of material handling, sewing and other processes involved in the conversion of fabrics into garments. This paper deals with qualitative and quantitative analyses of relationships between properties of apparel fabrics and garment making-up processes. Studies of mechanical properties of fabrics such as *extension, shear* and *bending* and their relationship to *objective tailorability* determinations constitute the main part of this research work.

INTRODUCTION

The Swedish apparel industry is gearing up its technical resources in order to meet the challenging demands of mechanization, automation and CAD/CAM application. Within the framework of a small, well defined programme of work named STRAPS — Swedish Technical Research in Apparel Production System, has been undertaken in some fields of apparel manufacturing technologies, especially those based on interface between material characteristics and processing techniques.

The programme of work under STRAPS is based on the following projects:

A. *Critical body measurements and computer-aided 2-D and 3-D pattern design*
The objective is to develop and adapt suitable technique for quick and accurate measurements of critical body dimensions, followed by using the measurements data for 2- and 3-dimensional garment designing and pattern construction (Figs 1, 2). Fabric properties such as extensibility, stiffness and draping shall be taken into consideration in CAD-geometrical design and shape.

B. *Material characterization in relation to garment making-up*
The objective is to characterize those material properties which are of importance as regards formability, tailorability, quality control and production processes. The work also includes analysis of physical and mechanical properties of fabrics which are impoortant for automation in apparel production processes.

C. *Development of new manufacturing techniques in garment production processes*
The objective is to evaluate and develop sensors and other equipment for facilitating automation and mechanization processes in some selected sewing procedures. Investigation on non-conventional shaping, sewing and joining procedures are also included in this project.

D. *Analysis of garment production system*
The objective is to evaluate the feasibility of various garment production systems based on process and material relevant data analysis. Investigations include the studies of logistics in material flow, operators and operations efficiency and tailorability.

The STRAPS programme was prepared under the direction of the R&D Advisory Committee of the Association of the Swedish Ready-made Garment Industry (KIF).

Mechanical properties of apparel fabrics

The mechanical properties of apparel fabrics are important from the point of view of stresses applied to fabrics in the making up processes including, load–extension, buckling, shearing, compression as well as physical changes in the fabrics that result from application of outside forces on the fabric in a garment during use due to gravity and by body fit and body movements.

Kawabata evaluation system

Together with Professor Niwa of the Nara Women's University, Kawabata has developed objective measurement techniques for fabric handling, prediction of tailorability and garment performance, etc. Their work and other work on the KES system has been well documented in a series of books under the general heading 'Objective Evaluation of Apparel Fabrics'. TEFO is one of many research institutes in Europe who are in possession of Kawabata instruments. These instruments are being extensively used for fabric characterization under the STRAPS-programme.

ANALYSIS AND PREDICTION OF TAILORABILITY

Very few studies on the quantitative analysis of tailorability, as defined by processing efficiency, have been reported in the literature. In all the methods used for producing garments there will be an interaction between particular method used and certain

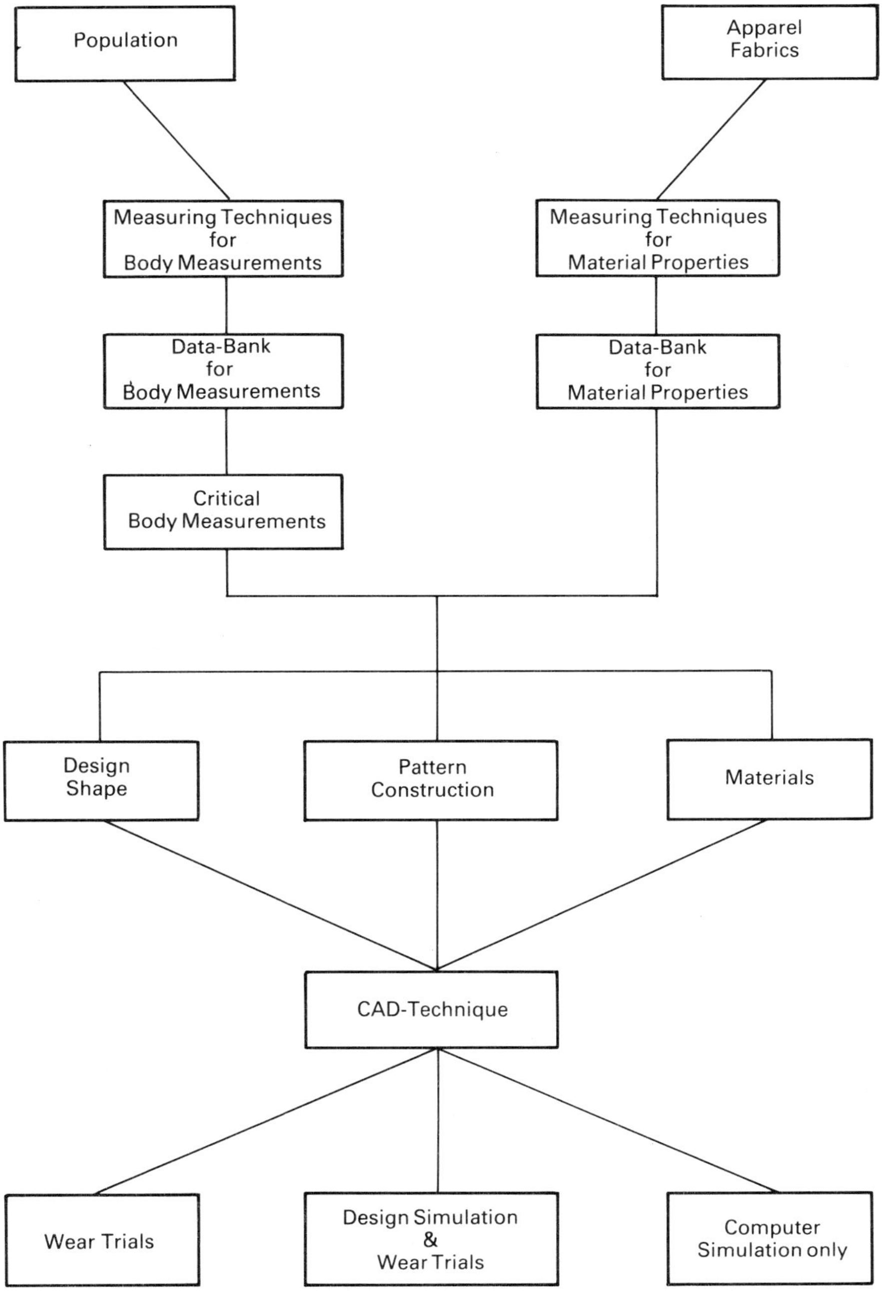
Population
Apparel Fabrics
Measuring Techniques for Body Measurements
Measuring Techniques for Material Properties
Data-Bank for Body Measurements
Data-Bank for Material Properties
Critical Body Measurements
Design Shape
Pattern Construction
Materials
CAD-Technique
Wear Trials
Design Simulation & Wear Trials
Computer Simulation only

Fig. 1.

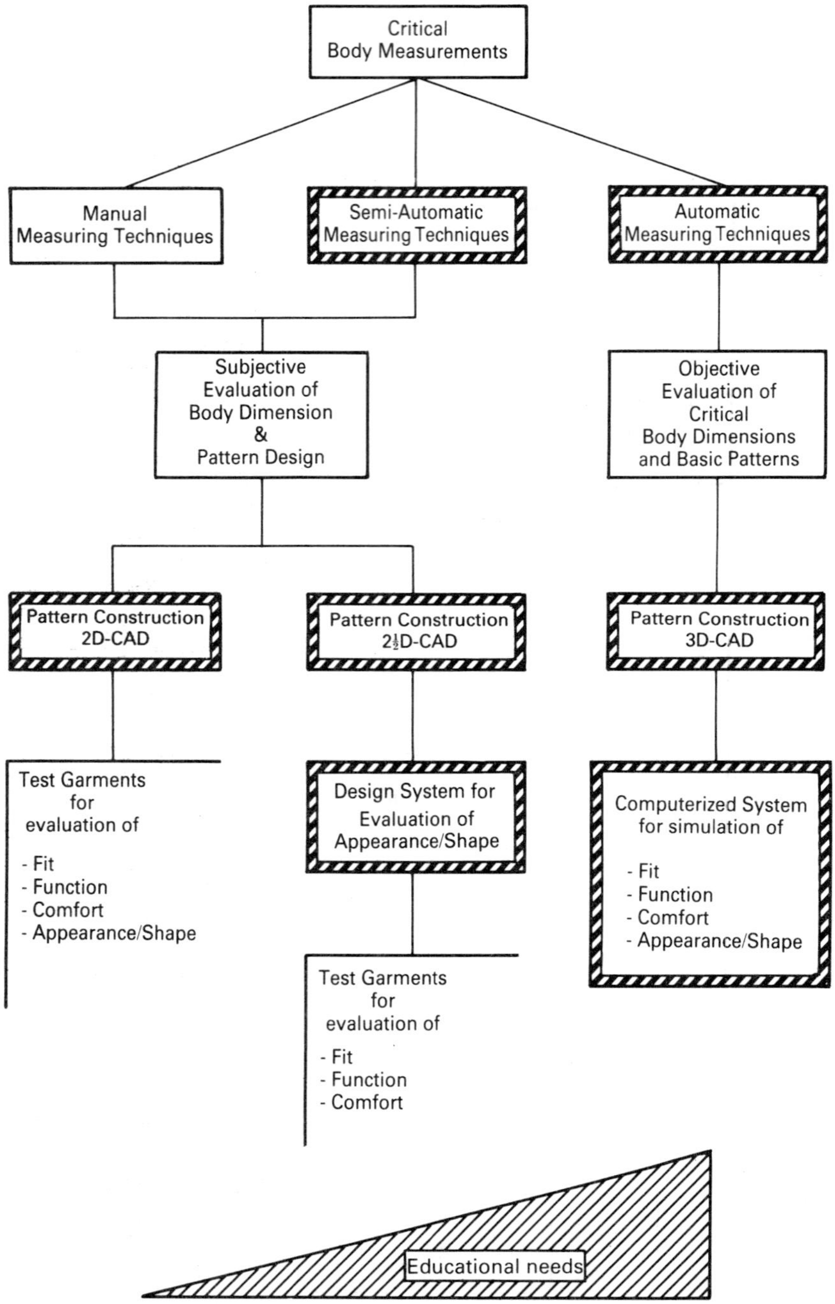
Critical
Body Measurements
Manual
Measuring Techniques
Semi-Automatic
Measuring Techniques
Automatic
Measuring Techniques
Subjective
Evaluation of
Body Dimension
&
Pattern Design
Objective
Evaluation of
Critical
Body Dimensions
and Basic Patterns
Pattern Construction
2D-CAD
Pattern Construction
2½D-CAD
Pattern Construction
3D-CAD
Test Garments
for
evaluation of
- Fit
- Function
- Comfort
- Appearance/Shape
Design System for
Evaluation of
Appearance/Shape
Computerized System
for simulation of
- Fit
- Function
- Comfort
- Appearance/Shape
Test Garments
for
evaluation of
- Fit
- Function
- Comfort
Educational needs

Fig. 2.

mechanical and physical properties of fabrics. Forming the fabrics with the techniques used, involves both uniaxial and biaxial extension and normal lateral compression of fabrics.

TEFOs production line

A test-line comprising machinery and process-control systems for the evaluation of different garment production systems, was installed at TEFO some years back. This line is schematically illustrated in Fig. 3, with description of various components.

Production Test Line

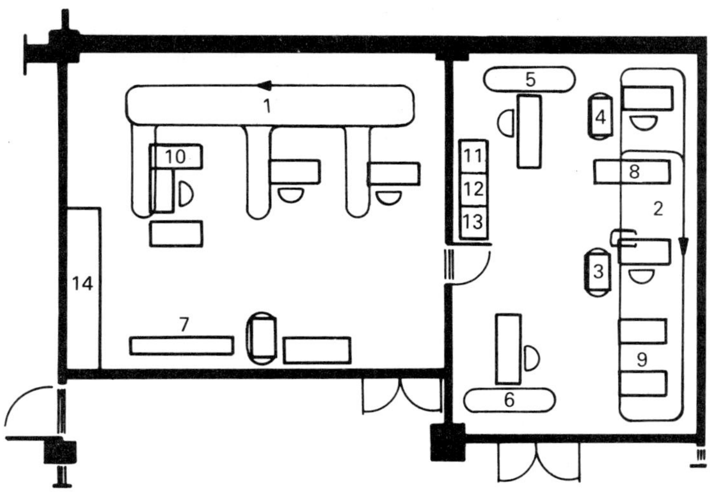

Fig. 3 — Schematic representation of TEFOs test line for garment construction. 1, Eton line system 2002 with data control. 2, Straight Eton component line (complete set of clamps). 3, RÅC system. Flexible layout for tests of different production philosophies. 4, Combined bundle-hanging system truck. 5, Eton work station (can be raised, lowered and extended). 6–8, Methods development station Eton 2002. 7–9, Straight system methods development station with extendable work station. 10, Planning board. 11, Production line for small parts make-up, such as bras etc. 12, Balancing system and Eton computers. 13, Piconet terminal. 14, GSD/CMA time setting system.

The installation contains among other things computerized systems for production data collection and standard time determination (CMA). This production line allows one to carry out full-scale evaluation of not only different production and processing units, but also material related processing efficiency, which TEFO has used for quantifying the tailorability. Production-line studies using a particular time and motion analysis, have been used as the basis for predicting how easy or difficult it would be to make a garment out of a certain material. The definition of tailorability not only covers the process of sewing, but also the processes of handling, feeding and transport of fabrics in garment making. By means of a finer analysis one can also study material relevant data for the control of each sub-process in garment manufacturing.

HYGRAL EXPANSION OF WOOL FABRICS

Reversible dimensional changes in wool fabrics due to humidity variations are important sources of fabric instability in garment structures in terms of unaccepted appearance in wear or after dry cleaning and repressing. Steam and chemical setting of wool fabrics enhance these changes and the dependence of hygral-expansion of the degree of fibre swelling during the setting process is well established.

The importance of the dimensional stability of woven wool fabrics and its effects on the garment making process has been pointed out by many workers and has been well summarized in an IWS technical bulletin by Shaw [1]. In tailorability prediction analysis work, it is necessary to include the aspects of dimensional stability of woven wool fabrics together with the mechanical and physical characteristics of material. Dimensional changes in apparel fabrics during pressing and wear, must constitute an integral part of basis for defining formability or tailorability equations of wool fabrics.

EXPERIMENTAL

Objective analysis of tailorability

TEFOs production test-line described earlier, has been used for studies on tailorability of apparel fabrics. The test garment type selected was a lady's blouse/jacket containing pockets and flaps. A large number of commercial test fabrics made of cotton, wool, polyester and blend fibres were obtained from the production line of some garment manufacturers.

The time taken for combined sewing and handling operations in the production of a test garment was registered on MTM time and motion studies.

For a given garment type and a given set of basic patterns, we believe that the tailorability of processed fabric can be quantified by standard time determinations using computerized method analysis, CMA. According to our hypothesis, CMA-time, obtain could be a good and practical measure of the ease of forming and tailoring flat fabrics into three-dimensional garments.

Measurement of fabric properties

The basic mechanical properties used in this study are those parameters defined and measured by KES-FB standard measuring system [2] as given below.

Tensile properties: EMT_1, EMT_2
Bending properties: B, $2HB$
Shearing properties: G, $2HG$, $2HG5$
Thickness

In general the measurements were carried out on fabrics conditioned in standard atmosphere of 65% RH and 21°C. In order to investigate the mechanical behaviour of wool fabrics at various hygral expansion levels, a series of measurements was also carried out on specimens after equilibrium with atmospheric conditions of 0% RH, 21°C and 90% RH, 21°C.

RESULTS

CMA-analysis and fabric mechanical properties

On the basis of analytical considerations, it is very likely that some interactions exist between KES-parameters measured of our test fabrics on one hand and the formability of tailorability of the fabrics on the other.

The mechanical properties of apparel fabric are important from the view of stresses applied to fabrics in making-up processes including, load-extension, buckling, shearing compression as well as physical changes in the fabrics that result from application of outside forces on the fabric in a garment during use due to gravity and by body fit and body movements.

Experiments concerning analysis of the model garment production process on TEFOs test line has given some interesting technical data. Fig. 4 shows CVMA-time

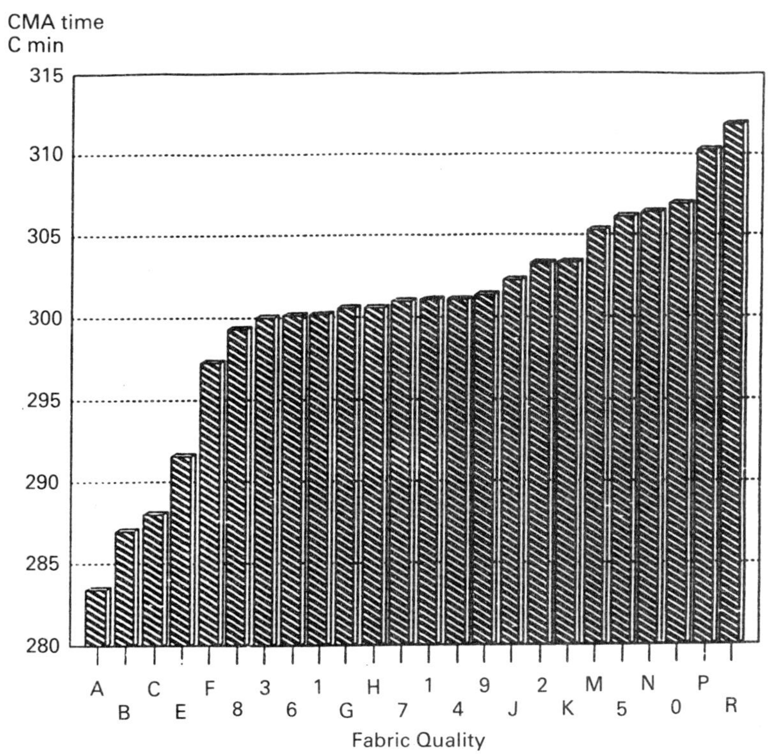

Fig. 4 — Computerized method-time analysis (CMA) of tailorability of 24 different fabrics.

distribution for 24 different fabrics made of varying fibre type and fabric structure.

CMA-time in this figure is the total garment production time and includes the processes of handling, feeding, adjustment, sewing, etc. It is the same garment style and set of basic patterns. The tailoring time varies from 283 C min for fabric A to 312 C min for fabric R. This sort of difference must be due to the direct influence of some physical and mechanical properties of the apparel fabrics being processed.

Extensive empirical analysis has been made at TEFO of the correlation between the CMA-times obtained for various test fabrics and relevant tailorability functions based on various mechanical parameters of these fabrics. Initially the CMA-results were plotted against the formability functions proposed by Lindberg [3] and Kawabata-Niwa.

$$F\ (\text{Lindberg}) = B \times EMT$$

$$F(\text{KN})\ (\text{eq. } 2) = \frac{EMT}{F_{\max} \times LT} \times \frac{B \times G}{2HG5}$$

No satisfactory correlation ($r = 0.53$) was obtained between Lindberg's formability value and CMA-time. Better correlation ($g = 0.76$) was obtained between Kawabata-Niwa's formability value and CMA-time (Figs 5, 6).

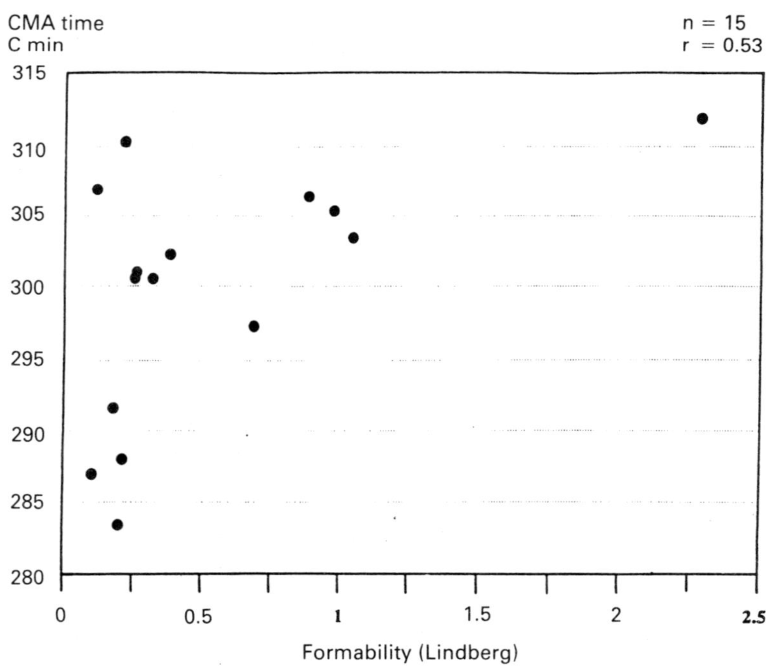

Fig. 5 — Correlation between total CMA-time and Lindberg's formability function.

Our further analysis of this work has shown, however, that good correlation exists between CMA-time and either log $(B \times EMT_1/2HG5)$ or LT log $(B \times EMT_1/2HG5)$, where EMT, is the level of extension of 50 g f derived from the

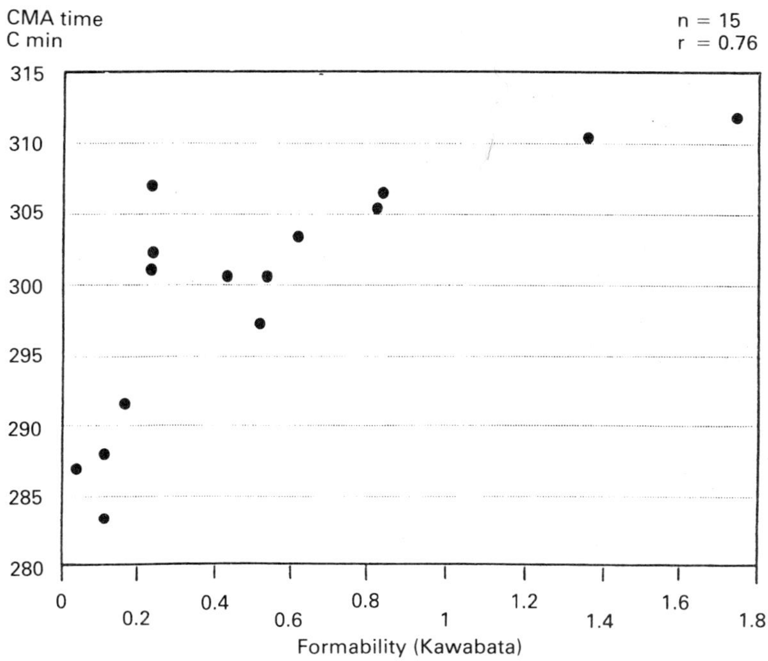

Fig. 6 — Correlationn between total CMA-time and Kawabata's formability function.

standard load–extension curve. As can be seen from Fig. 7, the correlation coefficient using our new function is 0.89. Figs 8 and 9 illustrate relationships between CMA and logarithmic functions of Lindberg and Kawabata-Niwa formability values [4,5].

Hygral expansion and fabric mechanical properties

Fig. 10 shows warpwise hygral expansion in two fabrics with increase in relative humidity. The level of hygral expansion is quite significant and the expansion shows a hysteresis effect due to reversal of humidity from 90% RH to 65% RH. For these fabrics there is a gradual reduction in thickness with increase of regain up to 90% RH and a slight hysteresis is due to humidity reduction from 90% RH to 65% RH (Fig. 11).

The effect of humidity change on some mechanical properties of the above mentioned fabrics are shown in Figs 10–12. As the relative humidity increases, the extensibility, *EMT*, decreases in warp and weft directions (Fig. 12).

The effect of humidity change on shearing properties of some wool fabrics is shown in Figs 13 and 14. There is an increase in both *G* and *2HG5* with increasing humidity. Hysteresis in shear *2HG5*, of these specimens remains nearly unchanged up to 65% RH but an increase in humidity from 65% RH to 90% RH causes a sharp rise in shear hysteresis.

Since the mechanical parameters, *EMT* and 2HG5, are present in TEFO's tailorability function, our results show that when the regain of a wool fabric is

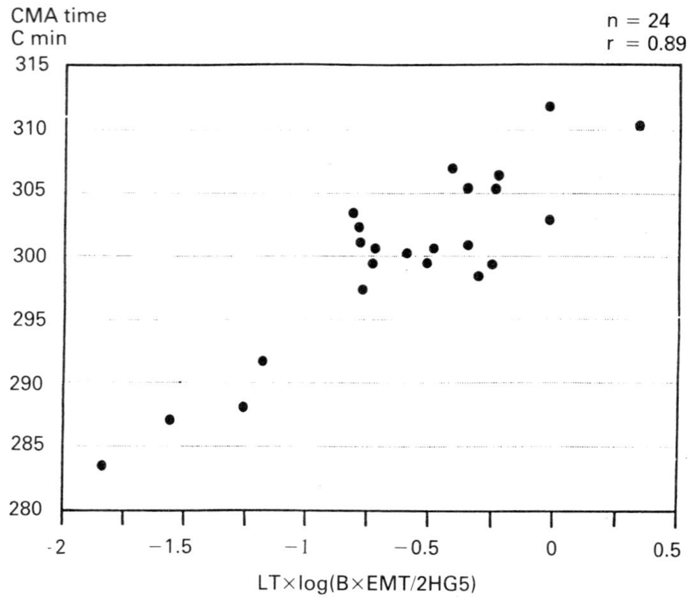

Fig. 7 — Correlation bbetween CMA-time and neew tailorability function (Shishoo and Choroszy).

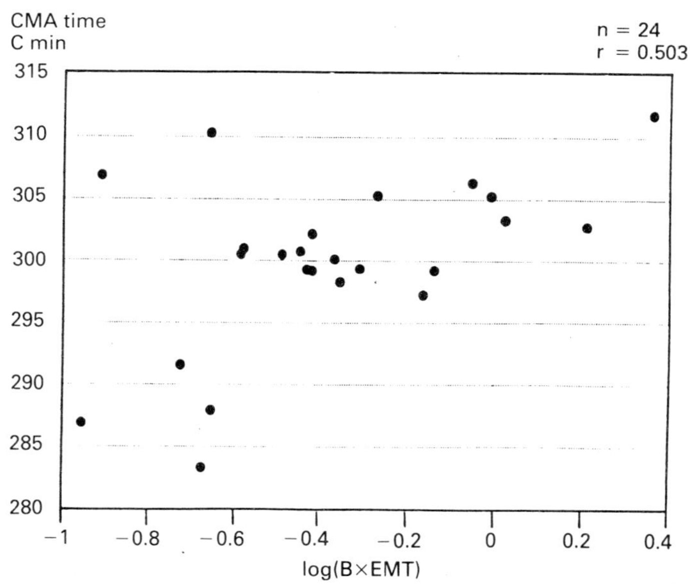

Fig. 8 — Correlatioon between CMA-time and log of Lindberg's formability function.

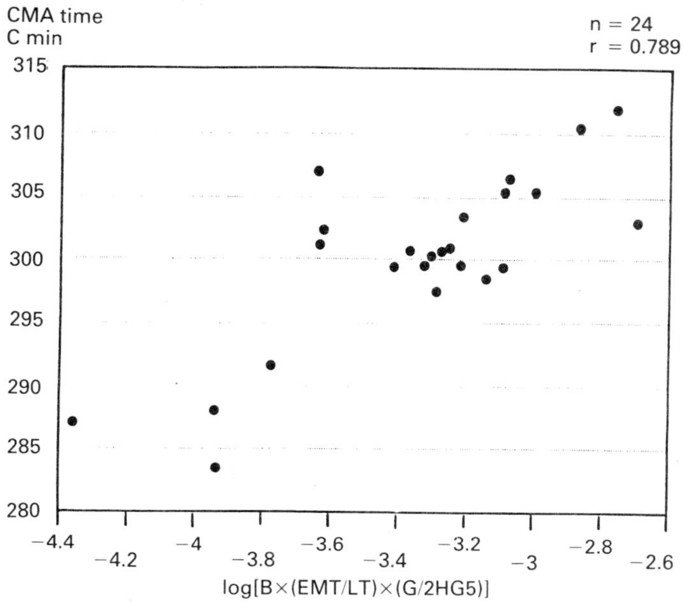

Fig. 9 — Correlation between CMA-time and log of Kawabata/Niwa's formability function.

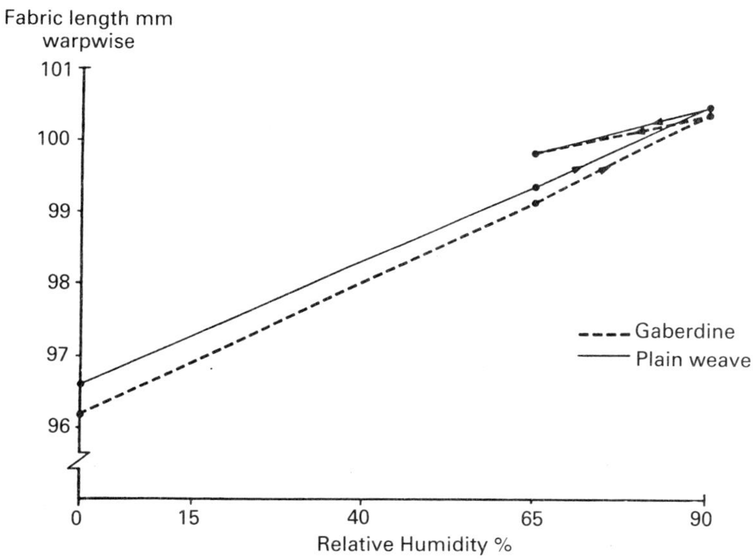

Fig. 10 — Dimensional changes in wool fabrics due to changes in relative humidity.

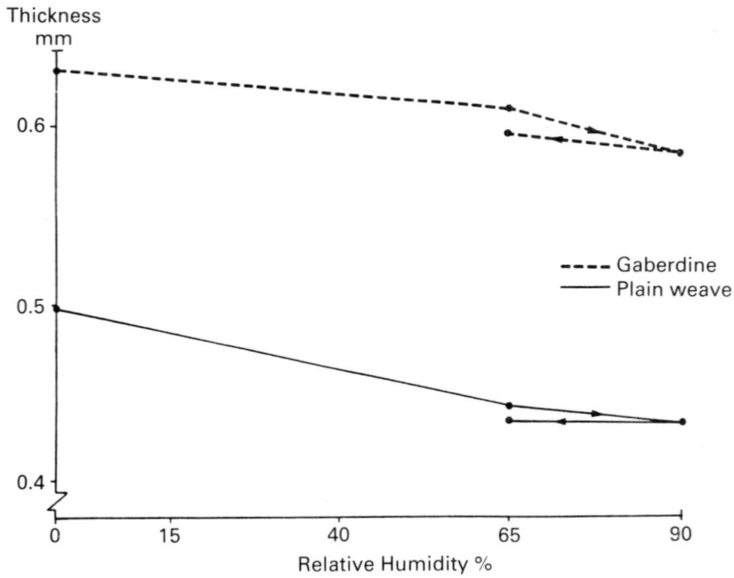

Fig. 11 — Thickness changes in wool fabrics due to changes in relative humidity.

changed in different stages of a garment making-up process, it will be accompanied by some important changes in those physical and mechanical prospects of the fabric which contribute towards its good tailorability. Therefore it is very important and necessary that in prediction models for tailorability one takes into account the aspects of hygral expansion and the hygral defendence of the critical fabric properties.

Fig. 15 shows that a linear relationship exists between hygral expansion and extensibility of wool fabrics. The tailorability requirements of a wool fabric demands a balance between these two parameters and form an important design criterion for the fabric manufacturer and the finisher. Fig. 16 shows the effect of steam- and thermal-press treatments on the extensibility parameter *EMT* of two wool fabrics. The measurements were made on specimens which were conditioned in atmosphere of 21°C, 65% RH after 10 repeated steam pressing and a single thermal pressing respectively. These results again point out the importance of testing and evaluating tailorability of wool fabric keeping in mind the variations in moisture, temperature and pressure which are very common in the making-up process [4,5].

Anthropometrical measurements and 2-D and 3-D pattern design

Within the STRAPS-project we are seeking a suitable stereo-photographic method which would make it possible to take various body measurements and allow quick and reliable handling of numerous measurements data. We are currently evaluating those critical body and garment dimensions which we believe are of fundamental importance for good garment-fit. Some of these critical body dimensions are at present not used because of the difficulty in their interpretation and use in conventional design procedures. However, we believe that the definition of critical body

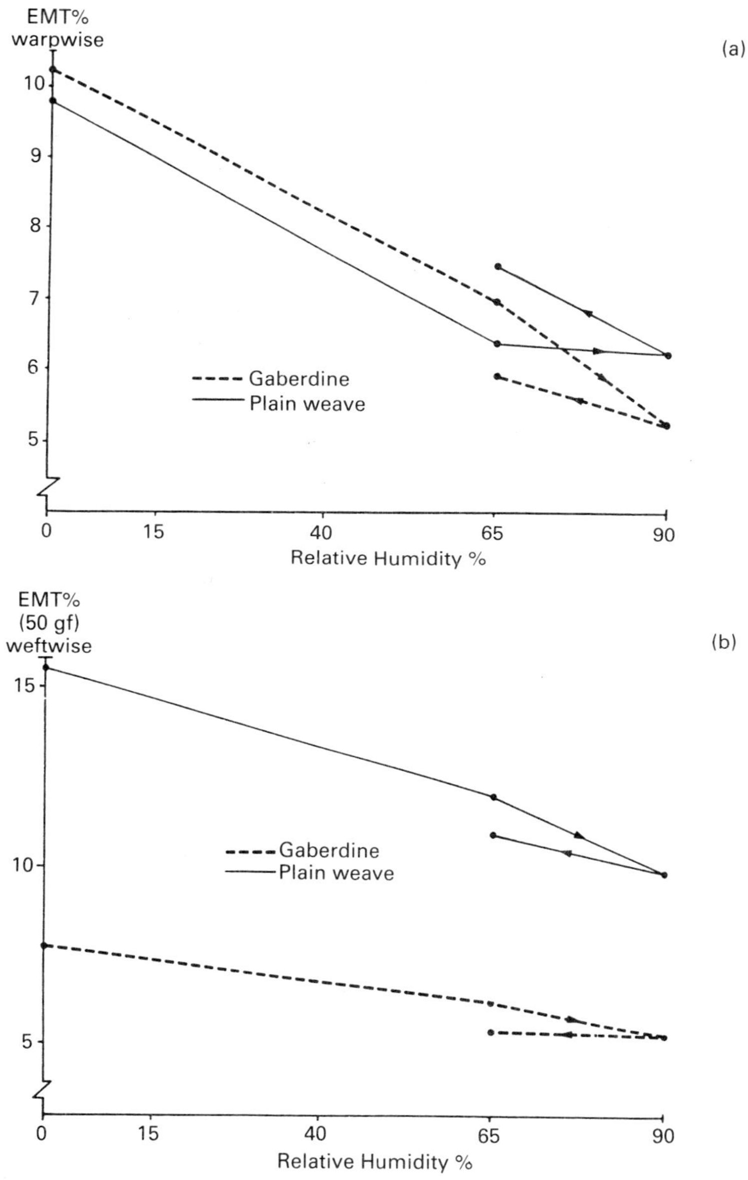

Fig. 12 — Effect of humidity changes on extensibility parameter, *EMT*, in warp (a) and weft (b) direction of wool fabrics.

measurements has to be renewed for making it technically more compatible with the theories and need of computerized designing of 2-D and 3-D patterns.

During the last decade or so CAD/CAM systems have been more used in the garment manufacturing industry. Computerized sketching and fashion illustrations have been available as a tool during the last 2–3 years but this technique is limited

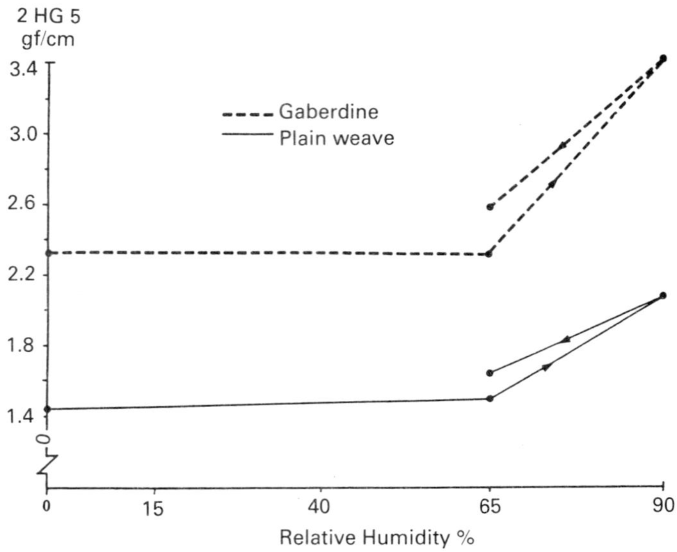

Fig. 13 — Effect of humidity change on shear hysteresis parameter, $2HG5$, of wool fabrics.

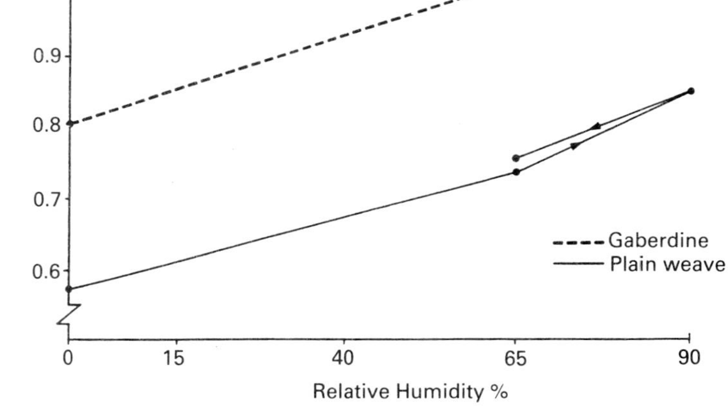

Fig. 14 — Effect of humidity change on shear stiffness, G, of wool fabrics.

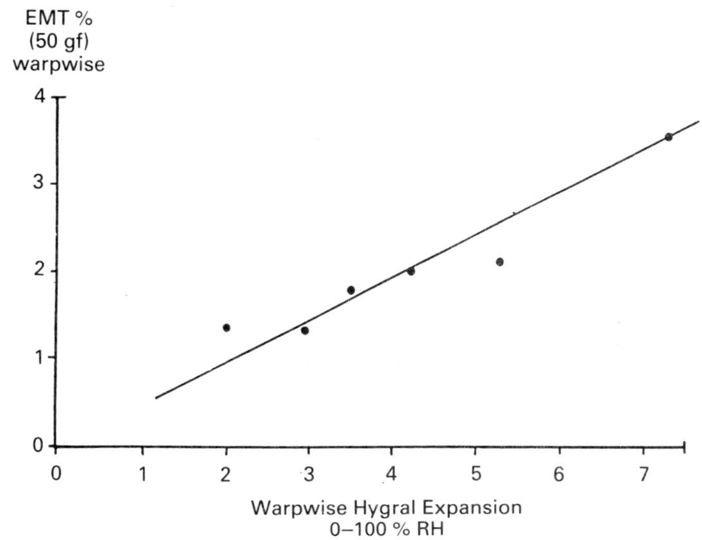

Fig. 15 — Relationship between extensibility parameter *EMT* and hygral expansion of six wool fabrics.

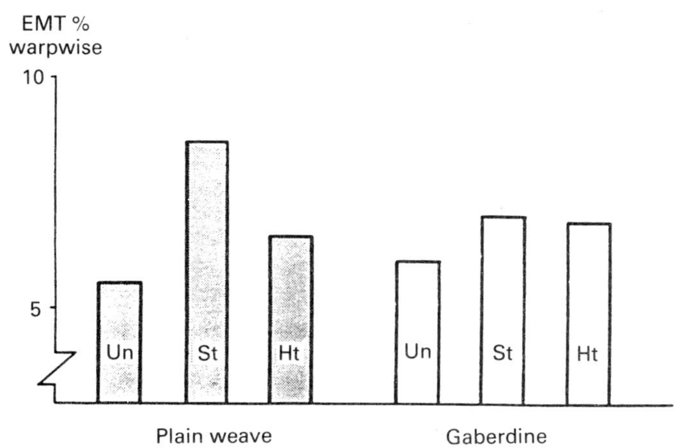

Fig. 16 — Effect of steam press (cycle of ten — 10 s press, 10 s vacuum) and thermal press (1 at 100°C 10 s, 10 s vacuum) treatments on extensibility parameter *EMT* of wool fabrics.

mainly because of the non-compatibility with the follow-up processes such as pattern design and pattern engineering. The existing CAD-systems are all very operator-controlled and characteristics processes are based on operators skill and experience.

Advantages of CAD-systems such as data-bank, quick response, material relevant garment design and flexibility in handling proposals of different garment styles are at present unknown to and unused by many designers. Through the STRAPS programme, we hope to develop new pattern designing principles and methods by making deeper analysis of material and anthropometrical data. Hopefully in the future 3-D pattern design systems it will be possible to simulate a CAD-created garment thereby eliminating the long process of making a large number of garment samples and evaluating them on dummies before the production process. Thus a designer would not only be able to evaluate how garment style is influenced by fabric characteristics but also how to manipulate basic patterns for fabrics with different mechanical properties in order to obtain one and the same garment style. Work is now progressing at TEFO with aim to establish quantitative relationships between fabric mechanical properties and garment design/shape. The experimental garment is a skirt based on a semi-circular pattern. Eight different fabrics have been selected and the test garments produced in two sizes. A test garment is then put on a dummy of corresponding size. Shape and draping behaviour is quantified. Studies are being made on the interaction between fabric mechanical properties and the garment shape with and without pattern manipulations.

CONCLUDING REMARKS

A new approach to study tailorability of apparel fabrics has been presented in this paper. This work is important from the point of view of automation in the garment industry as well as development work in wool fabric production. Some important formability/tailorability relevant fabric parameters have been measured at different levels of hygral expansion in wool fabrics, and their importance for the garment making process is discussed.

ACKNOWLEDGEMENTS

The author thanks Malgorzata Choroszy and Eva Tobisson for their skilful experimental work. Part of this work has been sponsored by the Wool Foundation and the Australian Wool Corporation.

REFERENCES

[1] T. Shaw, *Wool Sci. Rew.*, No. 55, 43 (1978).
[2] M. Niwa, T. Yamada and S. Kawabata,, Objective Evaluation of Apparel Fabrics. Textile Mach. Soc., Japan 67 (1983).
[3] K. Lindberg, L. Westerberg and R. Svensson, Wool Fabrics as Garment Construction Materials. *J. Textile Inst.*, No. 51 (1960), T 1475.
[4] R. Shishoo, Interactions Between Fabric Properties and Garment Making-up Processes. Annual World Conf. The Textile Institute. October 1989, p. 337.

[5] R. L. Shishoo and M. Choroszy, Analyses on Mechanical and Dimensional Properties of Wool Fabrics Relevant to Garment Making. The 8th International Wool Text. Res. Conf. Feb. 1990, Christchurch, N.Z.

QUESTIONS

DHINGRA: Regarding the relation between CMA and the various mechanical properties, I presume that you are trying to say that CMA can be a better measure of tailorability. Are we assuming that the final appearance of the garment you are making is exactly identical for all these fabrics?

SHISHOO: The method itself shows that there is no seam pucker in the same design. In the final appearance you assume that these fabrics appear identical when they are sewn together. I don't think that they will appear identical, partly because of the interaction between garment design and material properties that I have shown. If you take a skirt of the same basic pattern and you change the material the shape will be different. From the shape point of view they want to be the same. What I am saying is that we have to quantify what tailorability is. This is an effort from us to quantify the ease of production of a garment from a plain fabric to a 3-D construction. I cannot say that the appearance will be the same, in fact my results say that it will be different, because these materials are totally different but that is the reason that I am emphasizing the interaction of fabric properties and garment design.

SLATER:I have two questions. Firstly have you established reproducibility of the particular material by sewing several different samples to give you the same profile. Secondly which profile have you regarded as being the preferred one? What I am trying to establish is how much difference is there within samples and between samples if you like.

SHISHOO: As far as sewability is concerned we have four operators working. I don't expect very high correlation, but statistically the method is used in the clothing industry, and indeed in other industries. There will be some differences, because it is very difficult to reproduce fabrics, whatever method you use there will be (up to 10%) variation across the breadth of the fabric.

SLATER: What am I trying to establish is how realistic are the differences between the fabric parameters and how much of that difference is involved in the actual constructional variations.

SHISHOO: What I have shown is a very simple basic pattern.

SLATER: How many specimens have you sewn at each time? And how much variation was there between them?

SHISHOO: We have stitched four specimens at each time. There was very little variation. When we complete our quantification of the shape by using the Moire technique then we will see the variation, what I am saying is that we have to be very careful using the draping coefficient as something which is the only important parameter for design. Because we could have two skirts giving the same draping coefficient but a totally different appearance, so that quantification has to be made to make that clear.

AMIRBAYAT: I have two comments: you said someone established that extensibility and compressibility are the same.

SHISHOO: Not someone — Lindberg.

AMIRBAYAT: It is a rule of nature. A curve at a given point has one slope only. This is an established fact. Modulus under compression and modulus under tension at a given point is exactly the same, because the curve is continuous.

SHISHOO: We are dealing with non-elastic materials, what Lindberg said is that at smaller loads we can interpolate because if you go over from elastic to non-elastic region it is difficult to make any interpolation.

AMIRBAYAT: About hysteresis and shear modulus now at a 5° angle, the resistance against shear is mainly friction. So the higher the friction the higher the energy loss. Again this is fact; high friction, high modulus; therefore high friction, high energy loss so high hysteresis.

SHISHOO: It is not a fact.

GROSBERG: No, there is an elastic component in it. It all depends whether it is big or small. If the elastic component for the shear is big then what you say is not true. You should consider the elastic and the non-elastic components in shear hysteresis. The hysteresis is all loss but the slope of the curve depends upon whether there is an elastic component or not. It can be that there isn't. It depends on the curve, the total value, G, is all due to the hysteresis of the slope.

FAN: When you are quantifying the shape/drape of a skirt, on the same kind of fabric and pattern, what kind of differences have you found? Is there any difference in shape?

SHISHOO: There will perhaps be a difference. The integral part of our method is the use of a Moiré technique, we also have the draping coefficient. The area of shadow is not enough. We have to have more information and to quantify the shape, we start from a simple model which makes the calculation easier.

Introduction to Dr George Stylios

The next speaker needs little introduction, he is the organizer of this conference a task that for a day and a half performs very well. He did his M.Sc. and Ph.D. in the University of Leeds and came to Bradford four years ago where he is involved with research in the textile and clothing interface, studying problems associated with garment manufacture by employing advanced manufacturing technology to quantify and eliminate those problems. He is the Founder and Editor in Chief of the *International Journal of Clothing Science and Technology*. He is supported in this paper by Dr J. Fan, who had his B.Sc. from China Textile University and his Ph.D. at the University of Leeds. He has joined Dr Stylios as a postdoctoral research assistant at the University of Bradford. The title of the paper is 'An expert system for the prediction of fabric sewability and optimization of sewing and fabric conditions in garment manufacture', which is part of a Science and Engineering Research Council Funded research project.

An expert system for the prediction of fabric sewability and optimization of sewing and fabric conditions in garment manufacture (sewability system (SS))

G. Stylios and **J. Fan**
Department of Industrial Technology, University of Bradford, UK

ABSTRACT

Sewing problems are understandably determined by fabric properties and sewing conditions. This paper introduces one of the projects undergoing in the University of Bradford, which aims to develop an *expert system* for the prediction of *fabric sewability* and *optimization of sewing conditions* through *quantitative formulation* of the relationships between sewing problems and fabric, and sewing parameters. This approach is somewhat different to the general textile objective measurement of the KES and FAST, it concentrates upon specific sewing problems in garment manufacture. Sewability problems are predicted by *directly measuring* only those important properties related to sewability by means of easy to use *automatic testing modules* integrated with a personal computer, so that *no special knowledge or training* is required. A large part of this paper is devoted to describing the development of the system for the *prediction of seam pucker*.

INTRODUCTION

In the sewing stage of garment manufacture, seam pucker, i.e the deformation of fabric along the sewing line, and sewing damage, i.e. the damage of fabric in the stitch are two of the most common sewing problems which trouble the textile and clothing industries. Investigations into the causes of these problems [1–5] have revealed that these sewing problems are attributed, in general, to fabric properties, sewing conditions and mishandling. While mishandling can be avoided by skill or precaution, the solution to these problems is bound to be the selection or manufacture of the right fabric and the optimization of sewing conditions.

By fabric sewability, we mean the ability of fabric to be sewn without causing sewing problems, or in simple words, 'fabric properties related to sewing problems'. As far as seam pucker is concerned, where fabric is creased along the sewing line, sewing thread contraction, fabric thickness, flexural rigidity, Young's modulus, shear modulus, cloth tightness compressibility, etc., are known as important to a different degree on the occurrence of seam pucker. In the case of sewing damage, in which, fabric is damaged due to either the temperature build up in the needle during sewing, resulting in the melting of one or more yarns in the fabric or mechanical force acting on the fabric during the penetration and withdrawal of needle, resulting in the rupturing or cutting of one or more yarns, fibre properties, yarn strength, fabric compressibility, thickness, dyeing and finishing and the resistance of fabric to deformation can well be the cause of sewing damage.

With regard to sewing conditions, we include the selection of needle, sewing thread, thread tension, machine settings and the machine dynamics, which are inherently different not only between different sewing machines but amongst the same type and make of machinery. Here variables related to seam pucker are thread extensibility, thread stiffness, thread tension, presser foot pressure and sewing speed, and those related to sewing damage are needle coating, needle size, shape, thread surface properties, presser foot pressure, sewing speed and sewing room conditions.

Considering all these potentially relevant parameters to seam pucker and sewing damage, within which some are more important than others, and some can only take effect with the co-existence of others, we believe, the ultimate approach to either of these problems is to establish quantitatively the relationships between textile properties, fabric sewability and sewing conditions through extensive experimental investigations and to build these relationships into an expert system, so that industry can use it to predict sewability problems of supplied fabrics before they are cut or made up into a garment, to obtain advice to manufacture fabrics with less or no sewing problems and to eliminate sewing problems during garment manufacture.

Towards the above goal, the Sewability System (SS) for the prediction of seam pucker under optimized sewing conditions has been developed. In the following parts of this paper, the development and general feature of this system is presented.

SEWABILITY SYSTEM FOR PUCKER (SS)

This system is based on a mathematical model established from experimental data obtained from extensive investigations on commercial woven fabrics under industrial sewing conditions. The system is integrated with two automatic testing modules, which determine the thickness and the flexural rigidity of the test fabric. Unlike other systems, the subjectivity in placing the fabric for testing and reading of the results is eliminated to the minimum possible.

Mathematical model

The fabric data involved in the analysis for the establishment of the mathematical model, included the following fabric properties which are possibly relevant to seam pucker:

Thickness
Compressibility
Warp flexural rigidity
Warp frictional component in bending
Weight
Young's modulus
Shear modulus
Frictional component of shear
Air permeability
Warp density
Weft density
Warp count
Weft count
Cover factor
Weft flexural rigidity
Weft frictional component of bending

Excessive experiments of these fabrics, which are currently used in commercial garment manufacture, have been undertaken under industrial sewing conditions along the warp the weft and bias directions. Results along the warp direction were used, since sewing along the warp direction provided the most severe seam pucker. The seam pucker severity was assessed according to AATCC [6] standard. For the above fabric properties, some are more important in the prediction of seam pucker than others. The analysis began with establishing the most important properties related to seam pucker, which was found to be in turn: thickness, warp flexural rigidity, weft flexural rigidity, Young's modulus, shear modulus, frictional component of shear, warp density, compressibility, air permeability, warp count, and weft count. The relative importance of these fabric properties in causing seam pucker established by this analysis is in accordance with the mechanism of seam pucker due to structual jamming, one of the most difficult causes of seam pucker [3,5,7].

The SS uses the three most important fabric properties: thickness, warp flexural rigidity and weft flexural rigidity necessary for the prediction model, and the optional fabric constructional properties: warp density, weft density, warp count and weft count, necessary for the prediction model. Although fabric constructional properties are less important than Young's modulus, shear modulus, etc., they are usually available as part of fabric specification.

In order to assess the reliability of the mathematical model, the predicted pucker severity of sample of 50 representative fabrics were compared with their actual seam pucker severity occurred after sewing. No misprediction was found between unacceptable and acceptable seam pucker. Some misprediction only occurred between unacceptable and critical seam pucker and between critical and acceptable seam pucker.

Regarding the prediction performance of the SS when thickness, warp flexural rigidity and weft flexural rigidity are used in the prediction the correct prediction rate is 76%, when fabric constructional properties are also included in the analysis the correct prediction ratio is 81.4%.

System overview

With the seam pucker severity model established from the mathematical analysis discussed in the last section, a user friendly software was developed on a personal computer. The software was entirely written in the C computer language. The measurement of fabric thickness and flexural rigidity are carried out with two automated testing modules interfaced with the computer and integrated with the intelligent environment, these measurements are made automatically. It should be emphasized that handling of the test fabric and reading of the measured results is eliminated as much as possible and testing is done automatically. The user is given a choice of either having fabric constructional properties in the prediction or not. By entering fabric weight, and measuring fabric thickness and flexural rigidity in both directions, the likelihood of seam pucker occurrence is predicted with estimated possibility. If warp density, weft density, warp count, and weft count are used, the prediction accuracy is much higher. Fig. 1. shows the block diagram of the system and Fig. 2 shows a photograph of the SS.

The system contains an introductory file and an advisory file. The former provides users all essential information of this system and the latter advises the user how to circumvent the problem when it occurs.

FURTHER WORK AND CONCLUSIONS

Further work into optimization of sewing conditions is now under way. The most different hurdle that we try to overcome is variability of sewing machine tolerance; however, with advanced sensory techniques and the use of artificial intelligence we believe that these problems will be overcome. The clothing industry is one of the most labour-intensive and highly competitive industries. Without the right understanding of the relationship between garment and sewing conditions further development is impossible.

Under the mainstream of clothing automation, this project is concentrated on establishing quantitative relationships between sewing problems and fabric sewability and sewing conditions, then developing an automatic environment for the prediction of fabric sewability and optimization of sewing and fabric conditions and properties for garment and textile manufacture. The Sewability System (SS) developed for the prediction of seam pucker demonstrated the feasibility of a completely automatic intelligent environment. It is of particular importance that unlike KES and FAST which are used for general fabric classification, the SS demonstrates a different approach to predict and solve either through on-line changes of sewing conditions or properties of fabric during manufacture, specific sewability problems. Once all sewing problems are quantitativly formulated with fabric properties and

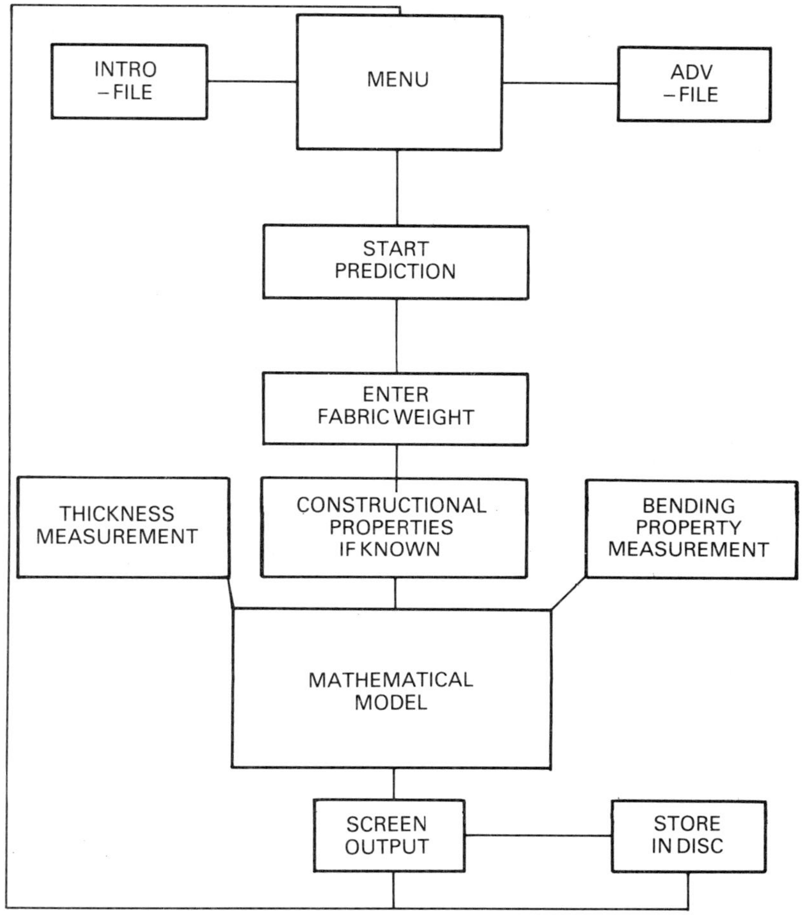

Fig. 1 — Block diagram of the Sewability System (SS) for the prediction of seam pucker.

Fig. 2 — The Sewability System (SS).

sewing conditions, this environment will be available. And this is believed to be the fundamental approach towards the automation of garment manufacture.

ACKNOWLEDGEMENTS

We would like to thank Mr H. Clay, Mr R. Deacon and Mr S. Bower for their assistance in the construction of the testing modules of the SSp.

REFERENCES

[1] Dorkin, C. M. C. and Chamberlain, N. H., Seam Pucker: Its Causes and Prevention, Clothing Institute Technical Report, No. 10, June, 1961.
[2] Dorkin, C. M. C. and Chamberlain, N. H., Seaming Damage and Pinholing in Cotton And Rayon Fabrics: Causes, Part 1, Seaming Damage, *Journal of the Textile Institute*, 1952, **43**, No. 5, pp. T203–233.
[3] Stylios, G. and Lloyd, D. W. 'The Mechanism of Seam Pucker due to Structural Jamming in Woven Textiles, *International Journal of Clothing Science and Technology*, **1**, No. 1, 1989.
[4] Stylios, G., Ph.D. thesis, The University of Leeds, 1987.
[5] Stylios, G. and Lloyd, D. W., 'Prediction of Seam Pucker in Garments by Measuring Fabric Mechanical Properties and Geometrical Relationships', *International Journal of Clothing Science and Technology,* **2**, No. 1, 1990.
[6] AATCC, 'Tentative Test Method 88B-1964T', *Technical Manual*, American Association of Textile Chemists and Colourists, 1964.
[7] Stylios, G. and Lloyd, D. W., 'A Technique for Identification of Seam Pucker due to Structural Jamming', *International Journal of Clothing Science and Technology*, *1*, No. 2, 1989.

QUESTIONS

JAYARAMAN: The key to the development of any expert system is the knowledge base. Is the knowledge base confined to those two properties and fabrics that you are using?

STYLIOS: We have more than 40 fabrics tested in more than sixteen properties. Now from the industry's point of view to do any realistic work regarding sewability problems you have to identify the range of fabrics that you have to work with. This group of fabrics are important. A lot of people say that they get seam pucker in suiting fabrics around the arm sleeves or in knitted fabrics; I do not call these problems seam pucker, I call them deformation due to uneven cut of fabric pieces and ropiness respectively. The fact is that to be realistic you have to look at it from a narrow angle of view and call in the industry, because at the end of the day fortunately or not the industry has to implement what we are doing, this is my own point of view.

QUESTION: Are you using patterened fabrics?

STYLIOS: No we are rejecting them at this stage.

COMMENT: Different patterned fabrics may influence the consistency of results.

STYLIOS: I agree and this is why we do not include them at this stage. This is a good point.

SARHADI: You have talked very little about the mathematical model, which is at the heart of the system.

STYLIOS: Two comments: Firstly there are industrial partners who should be the first to have access to the system, and secondly I am not a mathematical person myself. It takes a great deal of time to explain the mathematical principles of the model. We will be publishing papers about it in due course!

SARHADI: Is this making the system a suspect?

STYLIOS: I am not only myself alone in this. Dr Fan has done an excellent job and we are a good team.

JAFFER: What can mathematicians do with two factors, thickness and bending, except one linear multiple equation?

FAN: We have actually 16 fabric properties involved in the analysis, we have found the ranking of importance of these fabric properties: first thickness, then bending, then initial Young's modulus, etc., everybody is talking about formability which is based on two properties.

JAFFER: Yes, but two variables what mathematical model can you devise.

STYLIOS: The point that we are making is that the mathematical model is there and it is based on the 16 properties but if you look at the mechanism of seam pucker and rank these properties in relation to it, then you can identify that you have two most important properties. For the expert system to have a high probability of prediction you have to actually measure fabrics on these two important parameters only.

JAFFER: You did not mention that in the first place, in fact I have shown mathematically why thickness is important and I have quoted your data, I completely agree with you.

MUNDEN: I agree with everything that you have said, but it is the things that you have not said that are crucial as well, but let me say that the mathematical model is only going to determine seam pucker in woven fabrics due to structural jamming, when you talk about sewability you have to talk about the other aspects of it as well.

STYLIOS: This is a good point. This particular work concentrates on three sewing problems; the first is seam pucker where realistically we use the inherent problem of seam pucker, because if we try to include in this seam pucker due to differential feed for instance, you known as well as I do that a good mechanic can correct this cause of seam pucker, and if we include seam pucker due to handling or due to the extensibility of the sewing thread again this can be corrected. Therefore this work, which you see at its infancy, concentrates on what is the most important and realistic problem as far as seam pucker is concerned which is on the lightweight group of fabrics; the papery type of fabrics, the structually jammed fabrics if you like, and then we can go on to include the other aspects as well.

MUNDEN: I agree with you, I just wanted to clarify the fact that we were talking about sewability but seam pucker due to structural jamming.

STYLIOS:L As far as the particular system is concerned yes, but as far as the overview that I have given you together with the underlined principle of this project, it includes other sewing problems like seam damage, seam slippage, etc. One of the most important considerations is that the system is working and based on that, I hope we can incorporate the other problems as well.

MUNDEN: You have shown the mathematical model for seam pucker, but you have not yet shown the models for the other problems.

STYLIOS: No we are working on these right now.

MUNDEN: Can you please explain and I quote from you paper 'the prediction rate is 76%' what do you mean?

STYLIOS: With the present measurements based on 50 fabrics, 76 fabrics can be correctly predicted out of 100.

MUNDEN: So, 24 times out of 100 you are wrong.

STYLIOS: We are not trying to sell anything here, I am putting forward a line of work that has been done in a relatively short period of time, but let me say to you something else about expert systems, and people here that know about them would agree with me. If we can increase the fabric base that we have in this system, if we have a 76% correct prediction rate based on 50 fabrics or so, by increasing the fabric base to 200 fabrics we can have 100% correct prediction rate. The system was put together last week, and I felt that it was important to address the people in here exposing this important part of our work.

FAN: We are looking at an inherent fabric property which is related to seam pucker. On this principle if you put every fabric property into the system which we can do, then of course you get 100% prediction.

STYLIOS: If we had initial Young's modulus in this prediction, the prediction rate would have been 96%, but what we are trying to do is to tackle it from the point of view of the industry and say that here it is we simplified the prediction of seam pucker as much as we could and now we have to do more work to get the prediction rate higher.

SHISHOO: I think that next time we should arrange a seminar just on seam pucker, because as I pointed out in my lecture, we have different people working on the same topic using different properties for the prediction. I think this is an important topic and we have to find an ... tional agreement on which properties are important, eventually ... p in mind that there is an interaction between all those par ... to decide which of those fabric properties we use.

STYLIOS: I hope th ... take this work to the extent of ... other speakers to look t ... the principle of h ... now use exp ... v to do some ... esults by int ... u a line of ... to concent ... pucker results to

SHISHOO: I a ... on what parameters w

STYLIOS: Your poi

Session 3: Fabric sewability and garment manufacture

CHAIR: Professor P. Grosberg
Department of Textile Industries, The University of Leeds, UK

Introduction to Dr Harriet Meinander

This session is on fabric sewability and garment manufacture. Our first lecture this afternoon is on 'Seam damage predictions and KESF measurements', co-authored by A. Jylhä, E. Kelloniemi and Dr H. Meinander, who will be giving this lecture. Dr H. Meinander is at the Technical Research Centre of Finland; she obtained her first degree at Helsinki University and carried on to do a Ph.D. in textile and clothing technology. She has experience in many parts of the world, with IWS in the UK, and with the University of Stuttgart, as well as in quite a number of research places in Finland. I call Dr Meinander to give the next lecture.

Seam damage predictions and KESF measurements

Harriet Meinander, Aija Jylhä and **Eeva Kelloniemi**
Technical Research Centre of Finland (VTT), Tampere, Finland

ABSTRACT

Seam damage caused by the *needle penetrating* through the fabric can be a severe sewability problem, particularly when handling knitted materials. The risk for seam damage depends on fabric and sewing parameters: *fabric construction, fibre content, finishes, needle, sewing speed*, etc.

For the *prediction* of seam damage risk an L&M Sewability Tester was used. The penetration force transducer has been connected to a computer, which gives the possibility to study the *force curve* and the distribution of maximum values. The F-index was measured with a HATRA-sew device on a commercial sewing machine, and the number of damages studied in a microscope.

Seam damage is the result of a *high friction* between the needle and the fabric, either because of a tight fabric construction, *high fibre/needle friction* or *contamination of the needle* due to melting of fibres. By use of fabric softeners the risk of needle damage can be reduced.

The tightness of the fabric construction and the *surface friction* can be evaluated with the KESF *objective measurements*. Within a research project we studied the correlation between *sewability* and KESF parameters for a selection of knitted and woven materials, and the results of this study will be reported.

INTRODUCTION

Seam damage caused by the needle penetrating through the fabric can be a severe sewability problem. The extreme case is a knitted material with a tight structure and made of synthetic fibres, in which the friction between the needle and the fabric raises the needle temperature to values above the melting temperature of the fabric.

Eventually the needle melts small holes in the fabric structure, which in the case of broken yarns immediately or during use of the garment cause ladders in the knitted structure. The molten material also starts to stick on the needle, which increases the friction and seam damage tendency. In woven fabrics or non-thermoplastric structures, the problems are not as serious, but still seam damages reduce the value of the product.

The most important parameters, that have an infleunce on the seam damage tendency are:

— fabric construction (fibre content, yarn construction, structure, tightness, etc.),
— chemical treatments of the fabric (softeners, dyes, resin finishes, washing, etc.)
— needle thickness and design,
— sewing machine settings (presser foot design and pressure, sewing speed),
— sewing thread.

The needle, sewing machine settings and sewing thread must be chosen to harmonize with the fabric in order to minimize the risk for seam damage.

The central fabric properties, which have an influence on the seam damage tendency, are friction and tightness. As the KESF system is an excellent tool for measuring surface friction and mechanical properties of fabrics, a connection between the seam damage tendency and the KESF parameters was studied within a research project at VTT's Textile Laboratory.

SEWABILITY (SEAM DAMAGE) MEASUREMENTS [1]

For the prediction of seam damage three methods were used:

— needle penetration force, modified L&M Sewability tester,
— needle temperature rise, HATRA Sew measurements,
— subjective sewing test.

The L&M Sewability tester is a fairly simple device, in which a sewing needle makes a predetermined number of penetrations through the test fabric. The needle is attached to a force sensor. A high penetration force means a high resistance of the fabric and thus a high risk for damage.

In its original version, the device recorded the number of penetrations exceeding a certain limit, and this was used as a sewability index. We have connected the instrument to a computer, which registers the force peek of each penetration and calculates the mean force value, its variance and standard deviation. The distribution diagram of the force peeks and the penetration force curves of single penetrations can also be plotted (Figs 1 and 2).

The needle penetration action in the L&M tester is very slow (100 r/min) compared to an industrial sewing machine, and the test is normally run without

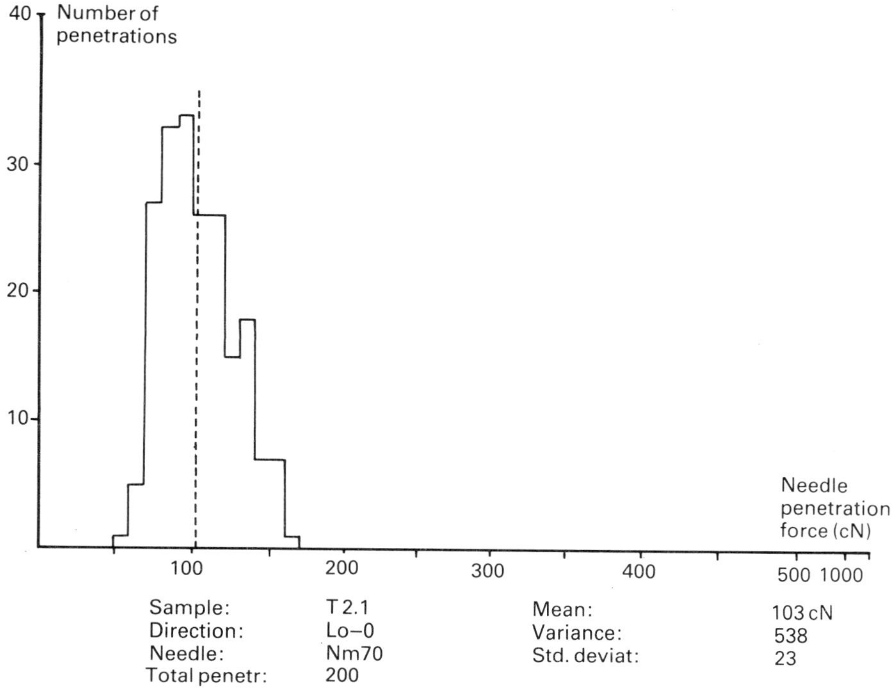

Fig. 1 — Distribution diagram of needle penetration force.

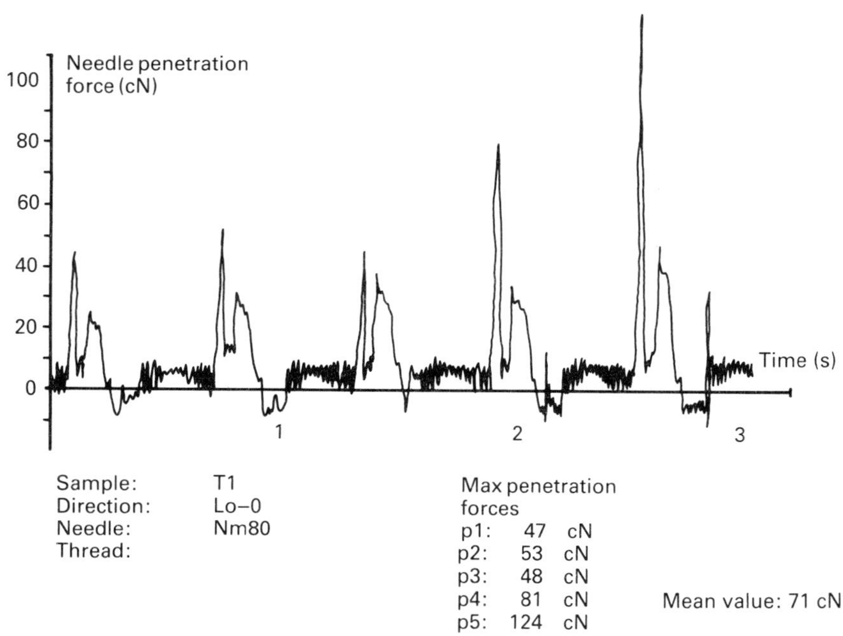

Fig. 2 — Penetration force curves, five penetrations.

sewing thread. In order to get a view of the behaviour in a real situation we therefore also perform a subjective sewing test on an industrial sewing machine with a maximum sewing speed of 5000 r/min (normally a plain lockstitch seam). The possible seam damages are studied with a stereo microscope and the gliding of the two fabrics relative to each other is measured.

The friction between the needle and the fabric is studied with the HATRA Sew device, which using an IR-sensor measures the temperature rise of the needle on the industrial sewing machine. A course needle (Nm 130) is used to in order to increase the temperature rise. The result is shown as an F-index 0 . . . 10, where 0 means no problem with temperature rise.

EXPERIMENTAL DATA

Two sets of fabrics were studied regarding their sewability (seam damage) and KESF properties (Table 1):

(A) woven sport wear fabrics ($n = 12$);
(B) weft knitted fabric for apparel ($n = 9$).

The complete series of KESF measurements was performed, as well as the L&M sewability, HATRA Sew and subjective sewing tests.

Table 2 shows the results from the sewability measurements. The woven fabrics distinctly fall into two categories regarding both the needle penetration force and the F-index: in sample 1–3 the values are very high, which clearly would indicate problems in the sewing operation. In samples 1 and 2 a number of damages could actually be studied in the fabrics.

For the knitted fabrics the results are not as clear. The measured needle penetration forces are relatively low, and no damages could be seen. Still the F-index values are remarkably high for some of the samples (6 and 9).

The results of the KESF measurements vary with the fabric type and direction, and as examples Figs 3 and 4 show the ranges and mean values of EM and B for the knitted and woven fabrics in warp and weft or wafe and course direction.

The relation between sewability and KESF parameters was studied by means of regression analysis of the results, Table 3. The KESF values are treated in three groups: resistance to dimensional changes (bending, compression, shearing and tensile), surface properties (friction and roughness), and construction (weight and thickness).

Generally there is a better correlation of the KESF values with the needle penetration force than with the HATRA Sew F-index.

For the woven fabrics a very strong correlation is found for two of the KESF parameters: $2HB$ and W. Not one of the surface properties show significant correlation with the sewability, nor with the gliding of the fabrics in the subjective sewing test.

Table 1 — Test materials

	Fibre	Structure	m^2-weight	Note
Woven materials for sports wear				
A1	100% PA	plain	140	Scotchgard
A2	100% PA	plain	140	Scotchgard
A3	100% PA	plain	140	Scotchgard
A4	100% PA	plain	105	Water-repellent
A5	100% PA	plain	105	Water-repellent
A6	100% PA	plain	105	Water-repellent, sanded
A7	100% PA	plain	105	Water-repellent, sanded
A8	100% PA	plain	105	Water-repellent, sanded
A9	75/25 CO/PA	plain	95	Scotchgard, sanded
A10	75/25 CO/PA	plain	95	Scotchgard, sanded
A11	75/25 CO/PA	plain	95	Scotchgard
A12	75/25 CO/PA	plain	95	Scotchgard
Weft knitted materials for apparel				
B1	50/50 CO/MO	plain	150	Ring sp. combed yarn
B2	50/50 CO/MD	plain	150	Rotor sp. combed yarn
B3	50/50 CO/MD	plain	150	Rotor sp. carded yarn
B4	100% CO	1 × 1 rib	175	Ion-free softener 5%
B5	100% CO	1 × 1 rib	175	Cation. softener 5%
B6	100% CO	1 × 1 rib	175	Cation. softener 5% Cation. dyefix. 3%
B7	100% CO	1 × 1 rib	175	Softener + stretch treatment 3%
B8	100% CO	1 × 1 rib	175	Stretch treatment 3%
B9	100% CO	1 × 1 rib	175	

For the knitted fabrics the shear properties (G, $2HG$ and $2HG5$) correlated well with the needle penetration force, particularly in the weft direction. On the other hand for the F-index, the best correlation is found for the bending properties in the warp direction (B and $2HB$).

DISCUSSION

The risk for seam damage in a fabric is a serious quality restriction, and should be evaluated together with the other fabric properties. The possibility to estimate seam damage risk from KESF parameters would bring an additional advantage to the Kawabata measurements.

Table 2 — Results from the sewability measurements

	Penetration force, cN (Nm7oR resp Nm80LAC)		F-index (Nm130)		Gliding (mm/70 cm)		Damages (1/10 cm)	
	Face side	Back side	Warp	Weft	Warp	Weft	Warp	Weft
A1	363	388	9.1	8.9	3	3	5	4
A2	310	334	9.0	8.9	1	3	2	2
A3	239	246	9.0	9.1	1	4	0	0
A4	26	29	0	0	2	2	0	0
A5	42	44	0	0.5	0	2	0	0
A6	12	14	0	0.1	4	1	0	0
A7	27	31	0	0.1	1	1	0	0
A8	17	18	0	1	0	1	0	0
A9	58	54	0	0	2	1	0	0
A10	59	54	0	0	2	1	0	0
A11	47	51	0.2	0.6	2	2	0	0
A12	52	51	0.2	0.2	2	4	0	0
B1	18	19	0	0	2	0.5	0	0
B2	31	32	0	0	3	1	0	0
B3	26	29	0	0	3	3.5	0	0
B4	19	17	4.0	2.6	2	2	0	0
B5	17	16	6.2	4.3	1.5	0	0	0
B6	25	23	8.2	7.4	3.5	2.5	0	0
B7	21	20	2.4	0.2	3	0.5	0	0
B8	15	15	1.2	0	1	0	0	0
B9	20	20	7.9	7.8	3.5	0.5	0	0

The test material in this study was limited, but already showed some direct correlations between sewability and KESF parameters. We will try to continue the work to develop a general model for the prediction of seam damage tendency of fabrics.

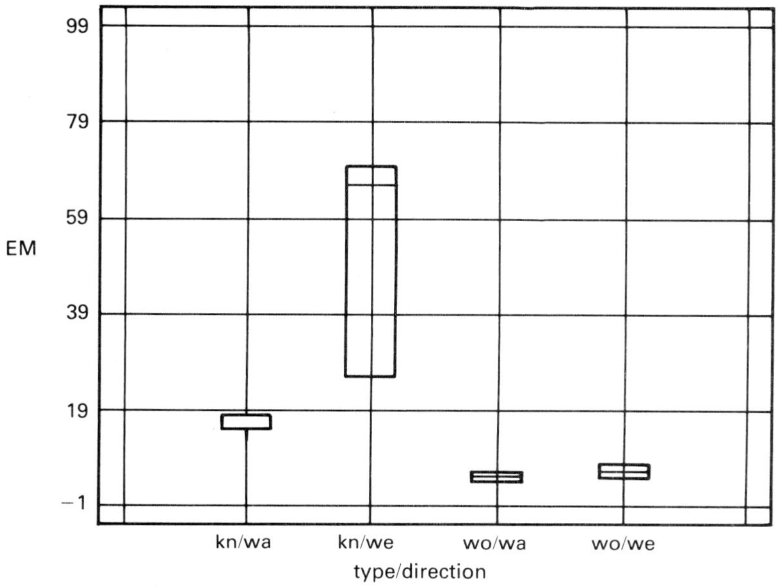

Fig. 3 — Ranges of *EM* values for knitted and woven materials.

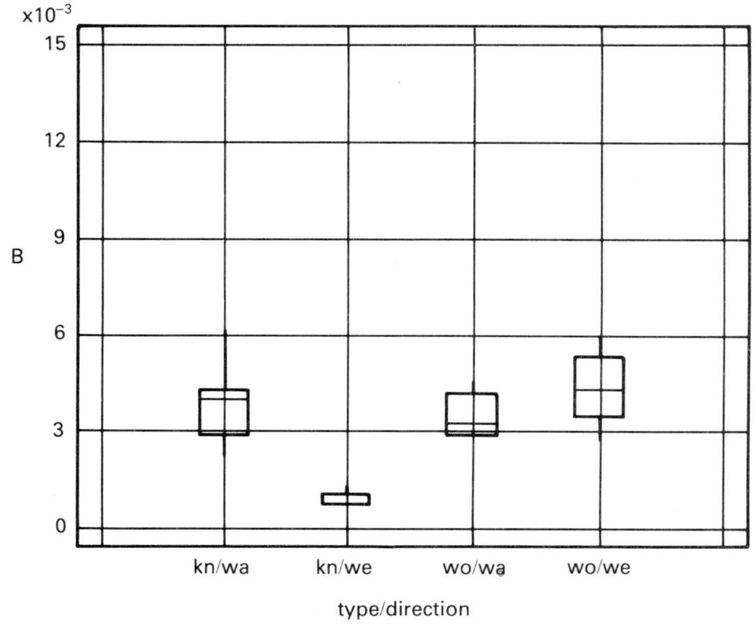

Fig. 4 — Ranges *B* values for knitted and woven materials.

Table 3 — Linear regression analysis, KESF vs sewability. Correlation coefficients

	F-index		Penetration force	
	Woven	Knitted	Woven (Nm70R)	Knitted (Nm80LAC)
LT,wa	0.102	0.641	0.151	0.817
LT,we	0.077	0.654	0.158	0.805
WT,wa	0.732	0.770	0.798	0.720
WT,we	0.406	0.672	0.247	0.680
RT,wa	0.812	0.176	0.947	0.352
RT,we	0.223	0.803	0.084	0.653
EM,wa	0.695	0.100	0.678	0.786
EM,we	0.749	0.670	0.372	0.826
G,wa	0.085	0.553	0.221	0.857
G,we	0.104	0.542	0.207	0.884
$2HG$,wa	0.256	0.470	0.070	0.864
$2HG$,we	0.475	0.134	0.253	0.917
$2HG5$,wa	0.412	0.367	0.632	0.596
$2HG5$,we	0.245	0.301	0.433	0.922
B,wa	0.432	0.885	0.702	0.546
B,we	0.273	0.122	0.206	0.606
$2HB$,wa	0.787	0.872	0.970	0.520
$2HB$,we	0.819	0.085	0.968	0.650
LC	0.209	0.022	0.020	0.016
WC	0.338	0.650	0.217	0.789
RC	0.070	0.302	0.078	0.519
EMC	0.012	0.517	0.149	0.650
MIU,wa	0.176	0.213	0.161	0.072
MIU,we	0.180	0.540	0.078	0.663
MMD,wa	0.666	0.441	0.742	0.756
MMD,we	0.316	0.208	0.104	0.673
SMD,wa	0.387	0.630	0.103	0.819
SMD,we	0.383	0.341	0.240	0.126
W	0.837	0.658	0.973	0.809
T_0	0.694	0.650	0.616	0.809

Wa: wale or warp direction.
We: course or weft direction.

REFERENCE

[1] Kelloniemi, E. and Jaala, U. Sewability of textiles and measurements of same. Technical Research Centre of Finland, Research Report 490, Espoo 1987. 87 pp. (in Finnish).

QUESTIONS

QUESTION: What sewing needles have you used for the needle penetration force measurement?

MEINANDER: We are using different needles for the knitted and woven fabrics; NM 70s for woven and NM 80s LSE for knitted. For the F-index measurements we used 130s. This is of course a coarse needle but it gives better differentiation between the fabrics.

STYLIOS: You have indicated that sewing damage depends upon the type of the needle. Can you specify what you mean by this statement?

MEINANDER: I mean the point of the needle.

DHINGRA: We have heard about sewability by various speakers and there was mention about hysteresis, bending, thickness and fabric weight. In your findings thickness and sewability gave you poor correlations, in an earlier presentation, thickness and bending was most important, can you please comment?

MEINANDER: We are discussing sewing damage, in the earlier presentation the focus was on seam pucker.

MUNDEN: You mention that 'the gliding of the two fabrics is measured'. What do you mean by the term 'gliding' and can you explain the relevant Figures 1, 2, 3 and 4?

MEINANDER: Gliding is the differential feed. In the figure we say 10 mm per 70 cm, so we stitch a piece of 70 cm and we measure in mm the differential feed.

MUNDEN: I am not surprised that one has difficulty in finding any correlations in the knitted, fabrics in terms of penetration force using the L and M Sewability Tester in my experience. I would not have thought that one can measure the force even if you have thousands of penetrations to within one unit, in fact they would all be low and cannot expect to find any damage with the force as low as that. So I think the lack of correlation that you have is because the knitted fabrics are so similar. If you are doing correlations you need to use different kinds of fabrics with different characteristics so you can get realistically different correlations.

MEINANDER: I agree the results are very close to each other.

STYLIOS: Just a comment on Professor Munden's point. Actually there are ways now to instrument sewing machines that can measure more than 300 points in a sewing cycle. You can get quite a lot of variation due to the various aspects of needle damage that we are talking about. Pioneering work at Denkerdorf and Leeds, which we now continue at Bradford, highlights how we can actually get more accurate results from the sewing machine itself in order to predict sewing damage.

MUNDEN: How many penetrations have you done?

MEINANDER: We do 200 penetrations in the warp and weft direction, and because there is not much difference we take the mean values.

MUNDEN: I cannot argue for the results on woven fabrics, because the tester was not devised to work with woven fabrics at all, but it was devised for knitted fabrics and certainly the variation in figures that you get over 200 penetrations have a significant difference of one.

SHISHOO: Two comments on what Munden said. First of all it is not true that there is no correlation at all in the set of those twelve samples because Dr Meinander was trying to see how the different finishing agents affect needle penetration. If you look at the table these are valid results; what is very important if you have to have experimental methods is whether they are able to show differences.

MUNDEN: What that proves is that those finishing treatments haven't affected the penetration force.

SHISHOO: This is an important finding. Secondly, yarn to yarn friction can be measured using the KES. Hysteresis in shear and bending are there and can be used for measuring any fabric, so we do not need to have a new method to measure yarn to yarn friction.

Introduction to Professor Sundaresan Jayaraman

You might have noticed that the next paper is also about sewing and on the 'Relationship between KES properties and sewing performance of difficult-to-sew woven fabrics'. This paper is by Professor Radhakrishnaiah and Dr Jayaramen. The paper will be given by Dr Jayaraman who is an Associate Professor at Georgia Institute of Technology in Atlanta, USA. He obtained his Ph.D. from the NCSU at Raleigh and his research interests are in CIM, KBS and structure/property relationships of yarns and fabrics.

Relationship between KES properties and sewing performance of difficult-to-sew woven fabrics

P. Radhakrishnaiah and **Sundaresan Jayaraman**
School of Textile & Fiber Engineering, Georgia Institute of Technology, Atlanta, USA

ABSTRACT

The *low-stress mechanical properties* of two groups of woven fabrics were evaluated using the Kawabata Evaluation System. The first group which consisted of *medium-weight denims* was produced by changing weaving process conditions including *loom type* (air-jet and flexible rapier), *shed geometry* and *yarn tension*. The variation in process conditions significantly altered the *low-stress mechanical properties* of these fabrics. The second group consisted of resin treated fabrics of different constructions.

The denim fabric which had a higher *tensile elongation* in the warp direction (when stressed to a peak level of 500 g/cm) gave the best performance in sewing. In the case of *resin-treated fabrics*, very high or very low values of elongation in one or both directions (warp/filling) caused noticeable deterioration in sewing performance. Improved sewing performance was associated with a range of *bending and shear properties*. This paper discusses the relationship between the low-stress mechanical properties of the two groups of fabrics and their subsequent *performance in cutting and sewing*.

1. INTRODUCTION

Tailoring a high quality suit is an art: it calls for a great deal of experience and skill. However, even the most experienced and skilled tailor may find it difficult to produce defect-free suits from fabrics that are not fit for suit production. Fabric properties, in addition to sewing skills, therefore, play an important role in the mass production of defect-free garments. In modern apparel plants, the assembly or joining of the component parts of a garment takes place at very high speeds. The standard operating time for many assembly operations in trouser manufacturing is of the order of seconds rather than minutes. This high-speed, high-volume environment also complicates the task of balancing production with quality; moreover, the importance of operator skills is substantially reduced under these conditions.

The increasing automation and mechanization of the apparel manufacturing process, therefore, requires accurate and precise control of process conditions (e.g. machinery speeds, settings and tension levels) based on fabric properties. A good understanding of the influence of fabric properties on the performance of production machinery and equipment is also very important to define policies and guidelines for the procurement of apparel fabrics. Proper regulation of the quality of incoming fabric materials, based on objective measurement of fabric properties, can eliminate several shopfloor problems faced by the apparel industry.

Fabric properties and sewing performance

In early research relating fabric properties and sewing performance, Lindberg and his associates have shown that the sewing performance of men's suiting materials depends on an index called 'formability' which is defined as the product of fabric bending rigidity and longitudinal compressibility [1]. Later, Spivak and Treloar showed that tensile and shear properties and longitudinal compressibility, measured at low load conditions in the two principal directions of the fabric, were related to tailoring performance [2].

Attempts to relate the low-stress mechanical properties evaluated in KESF (Kawabata Evaluation System for Fabrics) to sewing problems and the overall sewability of apparel fabrics began in the early 1980s. Since then, many Japanese and Australian scientists have reported on the progress achieved in terms of understanding the influence of KES properties on fabric sewability [3–9]. Most of this work, however, focused only on tensile and shear properties and was further confined to the production of men's suits from wool and wool-rich fabrics.

Need for the research

A lot of ground, however, still remains to be covered when it comes to a comprehensive understanding of the influence of low-stress mechanical properties and surface characteristics on the sewability of apparel fabrics. The present work attempts to relate KES properties to the cutting and sewing performance of two groups of non-wool type apparel fabrics, each of which consisted of fabrics classified as good and bad sewing materials. The work also reveals the importance of weaving process conditions for the production of heavy-weight fabrics that would deliver good cutting and sewing performance in apparel manufacturing.

2. INFLUENCE OF WEAVING PROCESS CONDITIONS ON THE LOW-STRESS MECHANICAL PROPERTIES AND SEWING PERFORMANCE OF DENIM FABRICS

Experimental work

In this study, three denim fabrics A, B and C were produced under three different weaving conditions. Fabrics A and B were produced on an air-jet loom under normal and modified loom settings while fabric C was produced on a rapier loom under normal working conditions. The following changes were incorporated on the air-jet loom to produce fabric B:

(a) Maximum reduction of warp tension.
(b) Increased staggering of heald frames.
(c) Assymetric shedding (formation of a non-symmetrical warp shed with the top layer of the shed tensioned considerably less than the bottom layer).

These changes were introduced to reduce the height, firmness and the definition of twill lines on fabric A, which had a very raspy feel as compared to fabric C. However, the modifications showed only a marginal effect on surface raspiness but altered the low-stress mechanical properties of the fabric to a very significant extent. The three fabrics were graded for their cutting and sewing performance by an apparel company and their low-stress mechanical properties were evaluated using KESF.

Table 1 lists the instrument settings used in the evaluation while Table 2 lists the

Table 1 — KES instrument settings for fabric testing

Compression	rate of compression	0.02 mm/s
	maximum force	50.0 gf/cm^2
	area compressed	2.0 cm^2/circle
Bending	rate of bending	0.5 cm^{-1}/s
	maximum curvature	$\pm$2.5 cm^{-1}
	sample size $L \times W$)	5 cm$\times$1 cm
Surface	rate of traverse	1 mm/s
	tension on sample	10 gf/cm
	normal force friction	50 gf
	contact force, roughness	10 gf
	distance measure	3 cm
Shear	rate of shearing	0.417 mm/s
	maximum shear angle	$\pm$1°
	tension on sample	10 gf/cm
	sample size ($L \times W$)	5 cm$\times$10 cm
Tensile	rate of extension	0.1 mm/s
	maximum tensile force	500 gf/cm
	sample size ($L \times W$)	5 cm$\times$20 cm

physical parameters involved in five KES tests and their units of measurement. The average values obtained for the low-stress mechanical properties of the experimental fabrics and the ratings established for their cutting and sewing performance are shown in Table 3.

Discussion

From Table 3, it can be seen that the loom modifications made the fabric more extensible (higher *EMT*) and less resilient (lower *RT*) to tensile load applied in the

Table 2 — An outline of key parameters involved in the KES tests

Nature of test	Physical parameter	Unit	Symbol
Tensile	Elongation	%	*EM*
	Linearity	—	*LT*
	Tensile energy	gf.cm/cm^2	*WT*
	Resilience	%	*RT*
Bending	Bending rigidity	gf.cm^2/cm	*B*
	Hysteresis	gf.cm/cm	2*HB*
Shearing	Shear stiffness	gf/cm degree	*G*
	Hysteresis at 0=0.5°	gf/cm	2*HG*
	Hysteresis at 0=5.0°	gf/cm	2*HG*5
Compression	Linearity	—	*LC*
	Compreesional energy	gf.cm/cm^2	*WC*
	Resilience	%	*RC*
Surface	Coefficient of friction	—	*MIU*
	Mean deviation of *MIU*	—	*MMD*
	Geometrical roughness	micrometre	*SMD*
Weight and thickness	Weight per unit area	mg/cm^2	*W*
	Thickness at 0.5 gf/cm^2	mm	*TO*

warp direction. The modifications also resulted in a 50% reduction in shear stiffness (*G*) and shear hysteresis (2*HG*). Bending rigidity (*B*) of the fabric increased in the warp direction due to slackening of warp sheet but the rigidity in the filling direction showed a corresponding decrease. The overall bending rigidity of fabric remained close to that of Fabric A. There was a reduction in the coefficient of surface friction (*MIU*), surface roughness (*SMD*), compressive resilience (*RC*), percentage thickness compression (*EMC*) and fabric thickness (*TO*). These changes in mechanical properties and surface characteristics resulted in a perceptible improvement in the cutting and sewing performance. Cutting blades retained their sharpness for longer periods while working on Fabric B. Improvement was also reported in the handling of fabric in the sewing department, and in working efficiency.

Most denims are medium- and heavy-weight fabrics and their tensile and shear properties are regarded as too stiff for cutting and sewing operations [4,7]. In the present study, a reduction in the tensile and shear stiffness properties contributed to improved performance in cutting and sewing. The reduction in fabric thickness (*TO*) and coefficient of surface friction (*MIU*) may also have had a positive influence on cutting and sewing performance. This work clearly shows that the process conditions employed in the weaving of medium and heavy weight fabrics can influence cutting and sewing performance; consequently, it can be reasonably expected that the hand quality of finished garments will also be affected [10].

Table 3 — Low-stress mechanical properties of fabrics representing loom variables

A.

<table>
<tr><th rowspan="3">Fabric</th><th rowspan="3">Fabric rating for cutting quality</th><th rowspan="3">Fabric rating for sewing quality</th><th colspan="10">Low-stress mechanical properties</th></tr>
<tr><th colspan="6">Tensile</th><th colspan="4">Shear</th></tr>
<tr><th colspan="2">WC</th><th colspan="2">RT</th><th colspan="2">EMT</th><th colspan="2">G</th><th colspan="2">2HG</th></tr>
<tr><th></th><th></th><th></th><th>W[a]</th><th>F[b]</th><th>W</th><th>F</th><th>W</th><th>F</th><th>W</th><th>F</th><th>W</th><th>F</th></tr>
<tr><td rowspan="2">A</td><td rowspan="2">Average</td><td rowspan="2">Average</td><td>17.8</td><td>7.0</td><td>45.1</td><td>44.6</td><td>3.1</td><td>1.3</td><td>7.3</td><td>7.5</td><td>14.7</td><td>15.0</td></tr>
<tr><td colspan="2">12.4</td><td colspan="2">44.9</td><td colspan="2">2.2</td><td colspan="2">7.4</td><td colspan="2">14.9</td></tr>
<tr><td rowspan="2">B</td><td rowspan="2">Good</td><td rowspan="2">Good</td><td>23.8</td><td>7.5</td><td>37.4</td><td>46.0</td><td>4.1</td><td>1.4</td><td>4.8</td><td>5.2</td><td>9.9</td><td>10.4</td></tr>
<tr><td colspan="2">15.6</td><td colspan="2">41.7</td><td colspan="2">2.2</td><td colspan="2">5.0</td><td colspan="2">10.2</td></tr>
<tr><td rowspan="2">C</td><td rowspan="2">Average</td><td rowspan="2">Average</td><td>15.7</td><td>8.9</td><td>41.7</td><td>42.1</td><td>2.7</td><td>1.7</td><td>7.6</td><td>6.4</td><td>15.4</td><td>12.9</td></tr>
<tr><td colspan="2">12.3</td><td colspan="2">41.9</td><td colspan="2">2.2</td><td colspan="2">7.0</td><td colspan="2">14.2</td></tr>
</table>

[a]Measured in the warp direction.
[b]Measured in the weft (filling) direction.

B.

Fabric	Fabric rating for cutting quality	Fabric rating for sewing quality	Low-stress mechanical properties						
			Bending	Surface		Compression			
			B W[a] F[b]	*MIU* W F	*SMD* W F	*WC*	*RC*	*EMC*	*TU*
A	Average	Average	1..01 3.81 2.41	0.209 0.215 0.212	7.1 10.4 8.75	0.118	40.3	22.3	1.249
B	Good	Good	1.59 3.35 2.47	0.185 0.210 0.198	6.4 9.8 8.10	0.104	34.6	19.1	1.178
C	Average	Average	1.88 3.99 2.94	0.207 0.212 0.209	6.3 12.3 9.3	0.118	41.3	20.6	1.220

[a]Measured in the warp direction.
[b]Measured in the weft (filling) direction.

3. SEWING PERFORMANCE AND KES PROPERTIES OF RESIN TREATED FABRICS

In this part of the research, the relationship between sewing performance and KES properties of resin-tested spun-viscose rayon fabrics is discussed. These fabrics represent a special group of apparel fabrics designed for use in the construction of armoury bags, which fit into the hollow barrels of guns and contain gunpowder. Four different fabrics, two of which were classified by the apparel manufacturer as good sewing materials (Red 1 and White 1) and the other two (Red 2 and White 2) as bad sewing materials were involved in this study. The fabrics and their characterization in terms of sewing performance are shown in Table 4. Table 5 gives the tensile properties of experimental fabrics while Table 6 presents shearing and bending properties.

Table 4 — Characterization of the sewing performance of experimental fabrics

Serial no.	Fabric identity	Characterization in terms of sewability
1	Red 1	Good
2	Red 2	Bad (too soft)
3	White 1	Good
4	White 2	Very bad (too stiff)

Discussion

As seen in Table 5, Red 2 and White 2 (fabrics regarded as bad sewing materials) show a greater imbalance in their warp and filling-way elongations (EMT). Red 2 has

Table 5 — Tensile properties of resin-treated fabrics

Sample	Tensile properties					
	WT		*RT*		*EMT*	
	Warp	Filling	Warp	Filling	Warp	Filling
Red 1 (good)	3.6	7.5	42.2	34.3	1.6	2.9
Red 2 (bad)	12.2	4.9	40.6	38.9	5.5	1.7
White 1 (good)	6.9	5.0	30.7	36.1	2.5	1.9
White 2 (very bad)	1.8	3.9	58.9	47.8	0.6	1.7

Table 6 — Shearing and bending properties of resin-treated fabrics

Sample	Shear properties				Bending propertiees			
	G		2HG		B		*2HB*	
	Warp	Filling	Warp	Filling	Warp	Filling	Warp	Filling
Red 1 (good)	0.3	1.6	0.6	3.1	0.8	0.88	0.78	0.76
Red 2 (bad)	0.08	0.34	0.18	0.7	0.22	1.2	0.11	1.2
White 1 (good)	1.8	2.3	3.3	4.8	1.4	1.7	0.65	0.82
White 2 (very good)	11.0	10.87	22.3	21.5	too stiff		too stiff	

a tensile elongation of 5.5% in the warp direction and 1.7% in the filling direction. The shear stiffness (G) and shear hysteresis ($2HG$) values of Red 2 are very low as compared to that of Red 1. The imbalance in tensile elongation capacity and the low shear stiffness properties of Red 2 can reasonably be expected to cause twisted and distorted seams and problems of shape formation and shape retention. In fact, the shopfloor problems encountered with Red 2 confirmed this observation (hypothesis).

White 2 shows very low values of tensile elongation (EMT) and very high values of bending and shear stiffness. The problems experienced with this fabric namely, excessive wastage, low working efficiency, excessive replacement of cutter blades and sewing needles and occurrence of needle chews may also be attributed to the above normal values for tensile, shear and bending stiffnesses.

Therefore, from the results obtained for Red 2 and White 2, we can conclude that there is a range within which tensile, shear and bending properties of fabrics should fall for good cutting and sewing performance. Too high or too low values of tensile elongation in one or both the principal directions of the fabric and too low or too high values of shear stiffness in one or both directions, may give rise to shaping problems and low production efficiencies. Careful adjustment of tension level and speeds may be required while working with fabrics such as Red 2, which is too sleazy. Low values of bending stiffness may not be detrimental to cutting and sewing performance. However, fabrics showing high bending rigidity and bending hysteresis values will not be desirable for good cutting and sewing. Control of yarn-to-yarn and fibre-to-fibre friction and the degree of mobility of warp and filling threads also appears to be important. Needle chews observed with White 2 can be attributed to excessive stiffness properties and reduced of the warp and filling threads.

4. CONCLUSIONS

1. Process conditions employed during weaving of medium- and heavy-weight fabrics significantly affect their low-stress mechanical properties and subsequent cutting and sewing performance.
2. Twill fabrics woven on two different looms may show differences in definition and firmness of twill lines which, in turn, relate to the degree of raspiness associated with twill constructions.
3. The low-stress mechanical properties measured in the warp and filling direction of medium-weight denims are influenced by the ratio of warp and filling yarn tensions.
4. The cutting and sewing performance of denim fabrics is influenced by their low-stress mechanical properties and surface characteristics (tensile, shear and bending stiffnesses, fabric thickness and coefficient of surface friction). The tensile, shear and bending stiffness of fabrics that perform well on the shopfloor fall within a range.
5. The occurrence of needle chews in resin treated fabrics may be related to the degree of mobility of warp and filling threads in the fabrics.

REFERENCES

[1] J. Lindberg, L. Waesterberg and R. Svenson, 'Wool Fabrics as Garment Construction Materials', *J. Text. Inst.*, (1960), **51**, T1475.

[2] S. M. Spivak and L. R. G. Treloar, 'The Behavior of Fabrics in Shear', *Text. Res. J.*, (1968), **38**, 963.

[3] T. J. Mahar, R. C. Dhingra and R. Postle, 'The Investigation and Objective Measurement of Fabric Mechanical and Physical Properties Relevant to Tailoring', in *Objective Specification of Fabric Quality, Mechanical Properties and Performance*, S. Kawabata, R. Postle, and M. Niwa (eds), Textile Machinery Company of Japan, Osaka, pp. 301–316, 1982.

[4] K. Ito, 'The Use of Objective Measurement of Fabric Mechanical Properties for Process and Quality Control in an Apparel Manufacturing Company', in *Objective Specification of Fabric Quality, Mechanical Properties and Performance*, S. Kawabata, R. Postle, and M. Niwa (eds), Textile Machinery Company of Japan, Osaka, pp. 331–338, 1982.

[5] M. Wemura, 'The Importance of Fabric Mechanical Property Measurement, for a Modern Mass Production System of Suit Manufacture, in *Objective Specification of Fabric Quality, Mechanical Properties and Performance*, S. Kawabata, R. Postle, and M. Niwa (eds), Textile Machinery Company of Japan, Osaka, pp. 339–346, 1982.

[6] M. Mori, 'Fabric Design and Production on the Basis of Objective Measurement of Fabric Mechanical Properties in Cooperation with Apparel Company Engineers', in *Objective Evaluation of Apparel Fabrics*, pp. 55–66, 1982.

[7] K. Ito, 'Process Control for Tailoring Based on Objective Data About Fabric Properties — Progress in the Last Year', in *Objective Evaluation of Apparel Fabrics*, pp. 89–96, 1982.

[8] R. L. Shishoo, 'Fabric Properties in Relation to Making-up and Performance of Garments', in *Objective Evaluation of Apparel Fabrics*, pp. 97–112, 1982.

[9] Y. Matsui, 'Fabric Finishing on the Basis of Objective Measurement of Fabric Mechanical Properties by Cooperation with Apparel Company Engineers', in *Objective Evaluation of Apparel Fabrics*, pp. 301–310, 1982.

[10] *The Standardization and Analysis of Hand Evaluation*, edited by S. Kawabata and published by the Textile Machinery Society of Japan, 1980.

QUESTIONS

QUESTION:
How do you set the tension low?

JAYARAMAN: We do not. We experimented with and came up with a low tension.

COMMENT: You are always to weave at low tension. They must have weaved with too high tension.

JAYARAMAN: Yes they were weaving with too high tension. I agree that this problem changes the character of the cloth.

SHISHOO: In your conclusion you have given several reasons, perhaps that statement can be applied to woven fabrics, it cannot be applied to knitted fabrics, are you generalizing.

JAYARAMAN: No, I am not generalizing and I do not have a lot of supporting data.

Introduction to Umberto Biglia

The last presentation today is about 'The prediction of garment appearance from measured fabric properties'. Mr U. Biglia will be presenting this talk. He is in charge of fabric quality at the Vitale Barberis Canonico Company in Italy. He is an ex-student of the University of Leeds and he has been working in industry ever since.

The prediction of garment appearance from measured fabric properties

U. Biglia
Vitale Barberis Canonico (VBC), Italy
A. F. Roczniok
CSIRO Division of Wool Technology, Ryde, NSW, Australia
C. Fassina
Gruppo Finanziario Tessile (GFT), Italy
N. G. Ly
CSIRO Division of Wool Technology, Ryde, NSW, Australia

ABSTRACT

A simple measurement of fabric *pressing performance* as indicated by *crease angle* has been developed and applied to the *prediction of appearance of wool and wool-blend tailored jackets*. Industrial and laboratory trails have shown that *crease angle* can be used in combination with *fabric formability* as measured by the FAST system to give an assessment of *garment appearance* of light-weight fabrics after they are pressed.

1. INTRODUCTION

Garment appearance is the major selling factor and retention of that appearance produces satisfied customers. Good garment appearance can be regarded as the end product of a number of factors including garment design, choice of fabric, tailoring skill and successful pressing. A highly skilled tailor, given sufficient time, can produce a garment of satisfactory appearance from almost any fabric. However, today, a garment manufacturer is required to produce a garment in minimum time for minimum cost. In these circumstances the choice of fabric plays a significant role in the success of the tailoring operation, during both sewing and pressing. The ability to predict a fabric's performance at each stage of manufacture will streamline production and assist in accurately costing the making up of that fabric.

The garment's appearance is largely imparted by the final pressing operation. It is therefore important to be able to predict the pressing performance of a fabric. Two types of problems can arise during pressing; embossing, where an imprint of the seam allowance or underlying garment structure becomes visible on the face fabric; and seam blowing, where the fabric does not hold a sharp crease and the seam or crease takes on an unpressed or 'blown' appearance.

This work examines the use of a simple measurement of fabric crease angle to predict the pressing performance with respect to seam blowing. This aspect of pressing performance has then been directly related to final garment appearance. As pressing performance has a greater influence on garment appearance for light- as compared to heavy-weight fabrics, other fabric properties which may influence tailorability have also been examined.

2. EXPERIMENTAL

Three trials are reported here; the first a laboratory trial sewing individual seams, the second with a highly automated large tailoring company and the third with a smaller tailoring company. For each trial the fabric mechanical properties were measured using the FAST equipment [5]. The garments or seams were pressed in the usual way and the resultant appearance judged subjectively. The laboratory trial relied on a panel of consumer judges while the two industrial trials relied on a panel of consumer judges while the two industrial trials involved expert judges from the participating companies.

Trial 1: Laboratory

This trial was carried out at CSIRO. The fabrics used were sourced from a number of mills. The fabric details are shown in Table 1. Fabrics were cut into strips to mimic jacket seams both in orientation and overfeed levels [4]. Each pair of strips was sewn together, the resulting seams were pressed as they would be a jacket, and a panel of three judges was asked to rank the seams with respect to appearance. The judges was asked to divide the appearance range into three levels, with a seam scoring 3 for acceptable, 2 for borderline and 1 for reject. When the results were combined, the seams received a score ranging from 9 down to 3 resulting in seven levels of appearance. On this seven level scale a score of 7–9 is acceptable, 5–6 borderline and 3–4 reject. The seams were also judged for appearance before pressing.

Trial 2: Large-scale tailoring company

This trial involved 43 fabrics which were tailored into men's suit jackets by a large-scale tailoring company. The details of the fabrics including their mechanical properties are shown in Table 2. The fabrics were sourced from a number of mills and the garments were constructed on the normal production line using fusible interlinings. After making up the jackets were pressed and their appearance assessed. A number of assessments were made over a period of some weeks, allowing the garments to fully recover from tailoring and pressing. Where assessments varied with time the final assessment of appearance is the one we have used. The garments were

Table 1 — Details of fabrics used in the laboratory sewing trial

ID	*F*-1	*F*-2	*E*-1	*E*-2	*RS*-1	*RS*-2	*HE*-1	*HE*-2	*B*-1	*B*-2	*G*	*T*	*W*	*A*	*APP*
0 1	0.27	0.27	2.4	2.6	4.9	1.6	4.7	2.5	5.0	5.5	31	0.34	143	0	5
2	0.44	0.61	2.1	5.5	1.6	4.2	2.5	6.0	8.9	5.3	34	0.38	165	2	6
3	0.37	0.78	2.9	8.6	1.0	−0.5	4.5	7.2	7.1	4.7	24	0.42	172	0	5
4	0.19	0.36	1.5	4.0	1.2	1.1	3.0	4.8	7.7	5.4	27	0.43	169	36	2
5	0.22	0.30	1.4	3.2	1.7	0.6	2.2	2.4	9.8	7.3	46	0.43	174	8	4
6	0.17	0.44	1.5	4.9	3.5	0.7	3.8	5.0	7.6	5.9	37	0.44	178	0	7
7	0.19	0.58	1.5	7.0	3.1	1.6	3.5	7.7	7.8	5.4	34	0.37	161	2	6
8	0.24	0.45	2.4	6.0	2.0	1.2	3.0	4.8	6.5	4.7	52	0.34	148	6	1
9	0.15	0.42	1.5	6.7	0.7	0.4	3.4	4.8	5.9	3.5	20	0.37	149	18	1
10	0.23	0.53	2.4	7.3	0.3	−0.5	3.8	5.2	6.2	3.9	22	0.36	153	23	1
11	0.14	0.37	1.3	5.6	0.7	0.0	3.1	4.5	6.8	3.8	23	0.36	148	40	1
12	0.14	0.42	1.4	7.0	0.7	0.1	3.2	4.8	6.4	3.4	22	0.36	149	17	1
13	0.13	0.33	1.1	5.1	1.6	−0.2	2.6	4.8	6.4	4.5	30	0.37	166	32	1
14	0.30	0.65	1.7	8.1	1.6	−0.1	4.0	6.1	7.8	4.4	23	0.37	172	9	7
15	0.26	0.52	2.0	5.9	2.2	0.6	4.1	4.9	7.6	4.9	31	0.41	178	12	6
16	0.22	0.53	2.1	7.0	0.8	0.3	4.0	5.6	6.8	4.2	21	0.37	152	39	2
17	0.19	0.29	2.1	5.1	2.5	2.9	3.4	4.5	5.6	3.3	30	0.34	154	2	4
18	0.18	0.27	1.6	3.6	1.6	1.6	2.7	4.7	6.6	5.0	32	0.36	151	0	4
19	0.25	0.46	2.4	6.0	0.8	0.0	3.3	4.7	6.2	4.5	27	0.40	169	33	1
20	0.51	0.32	3.1	2.1	0.5	1.2	2.5	2.7	6.6	6.8	45	0.37	162	5	2
21	0.36	0.62	1.9	5.7	0.8	0.9	3.9	5.2	8.0	4.9	21	0.29	161	10	5
22	0.34	0.34	2.7	2.7	1.5	1.3	3.1	2.9	7.5	7.1	37	0.43	164	5	3
23	0.16	0.32	2.0	4.8	2.0	1.1	2.9	4.8	4.5	3.8	31	0.36	148	0	2
24	0.22	0.30	2.1	5.3	2.0	1.9	3.5	3.5	6.1	3.7	34	0.40	157	16	1
25	0.18	0.38	1.6	4.8	1.0	−0.1	2.9	4.7	5.8	4.9	26	0.39	168	33	1

Definition of symbols: *F*, formability (mm^2); *E*, extension at 100 gf/cm (%); *RS*, relaxation shrinkage (%); *HE*, hygral expansion (%); *B*, bending rigidity (μN.m), *G*, shear rigidity (N/m); *T*, thickness at 2 gf/cm^2 (mm); *W*, weight in g/m^2; *A*, weft crease angle (°), *APP*, appearance rating, −1 or −2, denotes a measurement in the warp or weft direction.

Table 2 — Details of the 43 GFT fabrics

ID	*F*-1	*F*-2	*E*-1	*E*-2	*RS*-1	*RS*-2	*HE*-1	*HE*-2	*B*-1	*B*-2	*G*	*T*	*W*	*A*	*APP*
1	0.36	0.52	2.2	3.0	1.2	2.0	2.0	2.9	10.6	11.0	38	0.38	159	11	5
2	0.44	0.56	2.9	6.1	0.8	0.0	2.9	3.2	8.7	5.3	28	0.41	164	7	5
3	0.34	0.33	2.5	3.7	2.6	0.1	3.3	4.0	7.5	5.5	19	0.40	147	27	1
4	0.26	0.40	1.5	2.4	2.2	1.7	1.5	2.3	12.9	11.8	37	0.38	155	14	1
5	0.93	1.55	2.0	4.2	2.8	1.2	1.5	1.4	29.3	21.3	64	1.28	330	38	5
6	1.80	1.76	3.2	3.4	1.9	4.0	1.4	2.7	36.1	35.2	45	2.70	422	20	4
7	2.67	2.43	4.8	5.6	1.5	2.0	3.5	2.4	29.4	22.8	25	2.62	340	19	4
8	0.70	0.83	4.8	6.9	2.3	−0.4	5.3	5.1	8.1	7.6	23	0.44	189	2	4
9	0.74	0.69	4.2	4.8	0.4	2.5	3.9	4.2	11.3	8.2	21	0.54	187	1	5
10	0.55	0.59	3.1	4.6	1.5	0.8	3.3	3.4	8.9	7.2	22	0.40	167	28	1
11	0.21	0.28	2.0	2.4	2.5	1.3	1.5	1.4	7.3	7.4	38	0.35	151	28	1
12	0.25	0.16	2.0	2.3	2.3	1.9	1.6	1.3	7.2	5.4	38	0.35	148	38	1
13	0.40	0.36	1.9	3.0	1.1	1.1	1.8	2.3	12.7	6.6	27	0.53	191	37	1
14	0.50	0.29	2.2	2.3	1.2	0.8	1.6	1.3	13.8	8.6	31	0.55	188	36	1
15	0.44	0.21	2.2	2.5	0.9	0.4	1.8	1.6	12.9	7.9	29	0.53	187	38	4
16	2.22	0.29	6.9	3.0	1.9	0.9	6.6	3.0	8.1	5.3	20	0.64	183	19	4
17	0.77	0.68	3.4	4.8	1.3	−0.2	5.1	3.9	11.8	7.7	21	0.51	221	10	4
18	0.31	0.39	1.9	2.6	1.2	1.6	1.9	1.9	9.6	9.0	36	0.47	159	11	4
19	0.33	0.61	2.0	3.9	1.7	1.8	0.0	5.7	9.8	8.4	18	0.67	181	19	1
20	0.88	1.51	2.3	7.4	1.2	0.7	3.9	5.9	20.3	10.6	14	0.84	191	28	5
21	0.65	1.64	3.7	8.1	2.0	−0.1	4.3	6.4	8.9	9.9	12	0.74	186	13	5
22	0.97	0.89	6.0	6.0	3.3	3.3	6.4	4.8	9.3	8.8	18	0.88	188	10	5
23	0.47	0.32	2.1	2.9	1.2	0.8	1.6	1.7	12.8	7.0	28	0.54	191	28	1
24	0.53	0.94	4.0	7.8	1.5	−0.7	4.9	6.0	7.5	6.9	24	0.44	199	37	1
25	0.24	0.41	1.8	4.9	0.1	0.2	2.2	3.4	6.7	4.7	23	0.38	161	30	1
26	0.26	0.22	1.9	2.0	2.2	3.0	2.3	2.2	9.5	7.6	44	0.37	152	24	5
27	0.20	0.35	2.2	4.4	1.3	0.0	2.0	2.8	4.6	4.0	29	0.35	150	12	3
28	0.15	0.47	1.5	5.1	3.2	0.1	1.8	3.8	5.9	5.6	19	0.36	121	35	1
29	1.89	0.74	8.3	5.2	1.1	0.8	8.7	4.8	13.3	8.4	18	0.97	223	1	2
30	0.32	0.35	1.8	3.9	3.0	1.6	3.3	3.6	10.2	5.8	24	0.42	171	13	4
31	0.75	0.25	4.3	1.7	2.4	1.0	4.3	2.1	9.5	9.3	28	0.54	236	34	2
32	0.87	0.65	5.7	3.9	4.7	6.9	3.0	1.7	8.7	8.7	16	0.79	202	11	5
33	0.33	0.08	1.8	0.8	2.0	0.6	0.5	1.2	9.7	11.5	45	0.36	146	34	1
34	0.20	0.36	1.4	3.9	2.3	2.0	1.9	4.3	7.3	5.4	37	0.35	147	11	2
35	0.65	1.03	3.5	7.5	0.5	0.2	4.0	5.7	10.2	8.1	20	0.48	173	5	4
36	0.45	0.56	2.9	6.0	0.8	0.5	3.1	4.4	7.9	4.9	27	0.45	186	2	4
37	0.42	0.34	3.0	3.8	1.7	0.4	1.9	3.0	6.9	5.2	30	0.38	160	13	5
38	0.30	0.51	2.6	4.8	1.2	0.8	3.3	3.1	5.5	5.9	28	0.38	156	14	2
39	0.27	0.28	2.2	2.9	1.5	0.4	1.5	1.5	7.8	7.0	33	0.36	145	24	1
40	0.24	0.23	1.6	2.5	2.0	1.7	2.2	2.1	8.6	5.8	34	0.36	141	23	1
41	0.37	0.81	2.9	7.5	0.2	1.2	3.7	5.1	6.8	5.5	20	0.40	172	13	5
42	0.20	0.18	2.0	2.2	1.2	0.8	1.1	1.2	6.8	5.5	29	0.37	142	28	1
43	0.32	0.23	2.3	2.6	1.6	1.2	1.7	1.7	9.5	6.4	34	0.38	146	31	1

divided into five levels of appearance; acceptable, borderline or reject with a few garments ranked between acceptable and borderline and between borderline and reject.

Trial 3: Traditional tailoring company

This trial involved 19 fabrics which were tailored into men's suit jackets by a smaller more traditional tailor. The details of the fabrics are shown in Table 3. The fabrics were all sourced from a single company and the jackets were fully sewn using traditional tailoring techniques; for example canvas interfacings were sewn rather than fused to the face fabric. After making up, the garments were subjectively assessed for appearance. The garments were graded into three appearance levels; acceptable, borderline or reject.

Fabric properties

The wool and wool-rich blend fabrics used in the above trials ranged widely in weight and mechanical properties [5]. The properties shown in Tables 1–3 were measured using the FAST equipment. The FAST parameter of formability quoted in this paper is the product of bending rigidity and low load extension. This parameter was first introduced by Lindberg [1] and since then has been used by various other workers to assess fabric tailorability [2,3].

As well as the usual mechanical and physical properties, fabric crease angles were also measured in order to examine pressing performance. Small samples of each fabric were folded and a crease pressed using a standard cycle of steam and vacuum (30 seconds steam, 10 seconds vacuum) on a Hoffman press. The samples were trimmed back leaving one centimeter of fabric on one side of the fold. The creases were then allowed to recover for about 30 minutes and the angle of the crease measured. The angle was measured again after 24 hours recovery under standard conditions (65% RH, 20°C). An example of the sample and measurement angle is shown in Fig. 1. The crease angles measured in this way are also detailed in Tables 1–3.

3. RESULTS

Crease angle

The laboratory sewing trial included seams designed to mimic the shoulder, sleeve insert and long seams on a man's jacket. One of the most critical seams with regard to garment appearance is where the sleeve head is inserted into the armhole. The sleeve head is overfed into the armhole and then the excess fullness is pressed and moulded to give a smooth three-dimensional shape. Fig. 2 shows the seam appearance after pressing plotted against the crease angle. The symbols at each point represent the appearance of that seam before pressing. The most noticeable feature of the graph is the cut-off in crease angle value above which all the seams were judged to have a poor appearance. This indicates that for the fabric weight range used (140 g/m^2–180 g/m^2) a maximum crease angle of about 15° is allowable before the pressed seam appearance is judged unacceptable. This cut-off is observed for all of the seams constructed.

Table 3 — Details of the 19 VBC fabrics

ID	*F*-1	F-2	E-1	E-2	RS-1	RS-2	HE-1	HE-2	B-1	B-2	G	T	W	A	APP
1	0.62	0.68	3.4	4.7	0.8	1.2	3.3	5.0	8.2	6.6	32	0.43	202	15	5
2	0.36	0.42	1.2	2.7	2.4	1.2	1.3	2.1	10.2	8.1	45	0.42	181	40	1
3	0.38	0.78	2.9	8.6	1.0	0.5	4.6	7.2	7.1	4.7	25	0.42	172	4	5
4	0.19	0.37	1.6	4.0	1.2	1.1	3.0	4.7	7.8	5.5	27	0.43	169	5	5
5	0.16	0.32	2.0	4.7	2.0	1.5	2.8	4.7	4.6	3.8	31	0.37	148	15	3
6	0.35	0.35	2.8	3.5	1.5	1.3	3.2	2.8	7.6	7.2	37	0.42	165	10	5
7	0.20	0.35	2.2	4.4	1.3	0.0	2.0	2.8	4.6	4.0	29	0.35	150	12	3
8	0.17	0.25	0.9	3.3	1.1	2.0	2.2	3.3	12.5	4.8	30	0.42	179	32	1
9	0.38	0.53	1.9	5.3	2.0	2.3	4.0	4.7	10.5	5.0	19	0.45	179	15	5
10	0.34	0.51	2.0	4.5	1.4	3.3	3.6	4.7	9.9	5.5	22	0.46	183	13	5
11	0.41	0.54	2.0	5.1	1.4	3.4	3.5	5.2	10.6	5.1	21	0.50	181	13	5
12	0.25	0.32	1.4	3.7	1.1	1.2	3.3	3.3	12.1	5.1	29	0.44	179	18	3
13	0.27	0.41	1.4	3.8	—	—	—	—	11.9	5.5	27	0.45	182	26	1
14	0.15	0.27	2.0	3.9	2.8	3.1	3.0	4.0	4.4	3.7	34	0.34	150	19	1
15	0.10	0.23	1.4	3.7	2.7	1.3	3.5	4.3	4.6	3.8	39	0.31	148	9	3
16	0.24	0.40	3.0	5.4	1.9	1.5	4.2	4.7	4.2	3.8	26	0.35	153	6	5
17	0.30	0.46	1.6	4.5	1.4	2.2	3.1	4.0	9.5	5.2	27	0.38	187	17	3
18	0.30	0.42	2.7	4.6	2.9	3.4	3.6	5.5	5.7	5.1	28	—	148	2	5
19	0.23	0.27	2.3	3.5	1.7	4.1	3.6	4.1	6.2	4.8	29	—	135	4	5

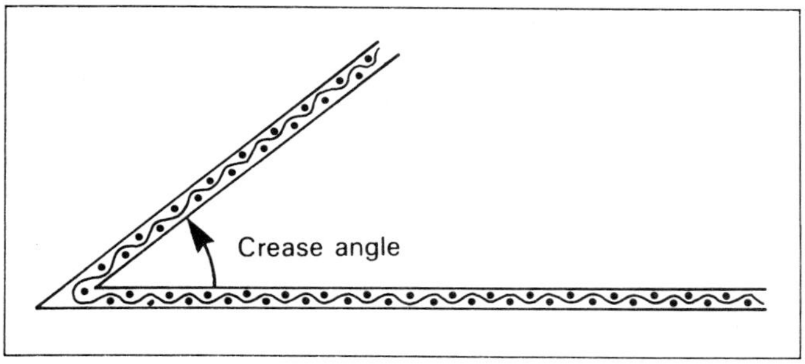

Fig. 1 — Crease angle sample and measured angle.

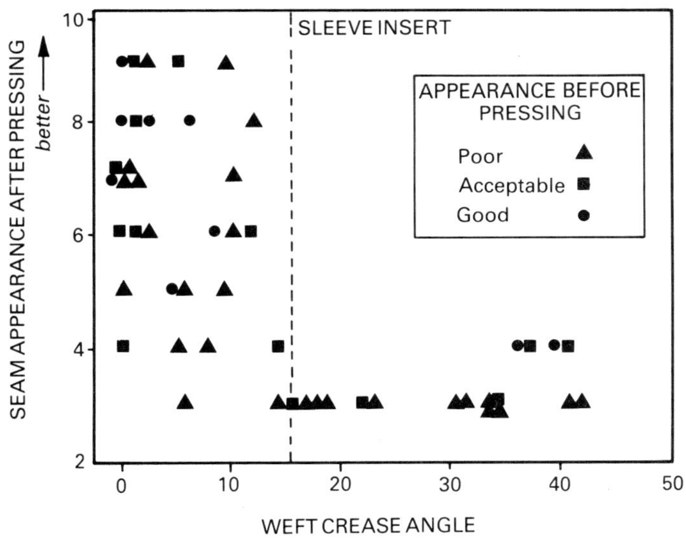

Fig. 2 — Trial 1 — Appearance vs crease angle.

The first of the industrial trials, conducted by a large garment manufacturer, resulted in suit jackets ranked by industrial experts into five appearance levels. Because of the wide range in fabric weight the same cut-off in crease angle is not apparent. If we consider only the fabrics with weights in the same range as the laboratory trial (i.e. between 140 and 180 g/m^2) a sharp cut-off in crease angle value is observed, at about 15°, above which all suits are rejected. The one exception to this, at a crease angle of 24°, was a more casual, younger style jacket which was accepted despite its poorer appearance. These appearance values are plotted, in Fig. 3, against the fabric crease angle. This trial shows that a low crease angle is a necessary but not

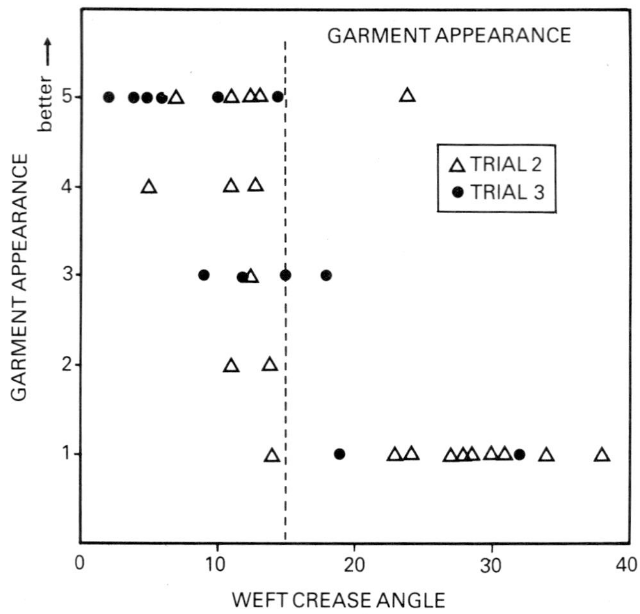

Fig. 3 — Trials 2 and 3 — Appearance vs crease angle.

sufficient condition for a good pressed seam appearance. In other words some of the fabrics which have a low crease angle still produce a garment of poor appearance.

The second industrial trial, conducted by a smaller tailoring company, using traditional construction techniques produced jackets judged to lie in three appearance levels. Again the wide range of fabric weights masks the cut-off in crease angle. When only the fabrics in the range 140 g/m^2 to 180 g/m^2 are considered, the cut-off in crease angle value again becomes apparent at about 15°. These results are plotted against crease angle in Fig. 3. As with the first trial, a low crease angle value did not guarantee an acceptable pressed appearance.

Mechanical properties

Fabrics with a low crease angle value did not necessarily produce acceptable garments. It was therefore important to determine whether or not poor sewing was the reason for rejecting the garments. The appearance of the seam before pressing can be regarded as a measure of the success of the sewing operation. In the laboratory trial, seam appearance was rated before as well as after pressing. The different symbols in Fig. 2 denote the appearance of the seams before pressing. If we examine the seams with a low crease angle, those with a poor pressed appearance do not necessarily have a poor appearance before pressing. This implies that there is more to producing an acceptable seam than sewing it well in a fabric with a low crease angle. Other fabric mechanical and dimensional stability properties affect how easily a seam is moulded into a smooth three-dimensional shape by pressing [4].

FAST data from the two industrial trails were examined in combination with the crease angle. The combination of crease angle with fabric weight or with warp extensibility ($E100$) correlated well with the final appearance rating, but the best fit was always found for the combination of crease angle with warp formability. This relationship is shown in Figs 4, 5 and 6. The symbols denote the appearance rating of

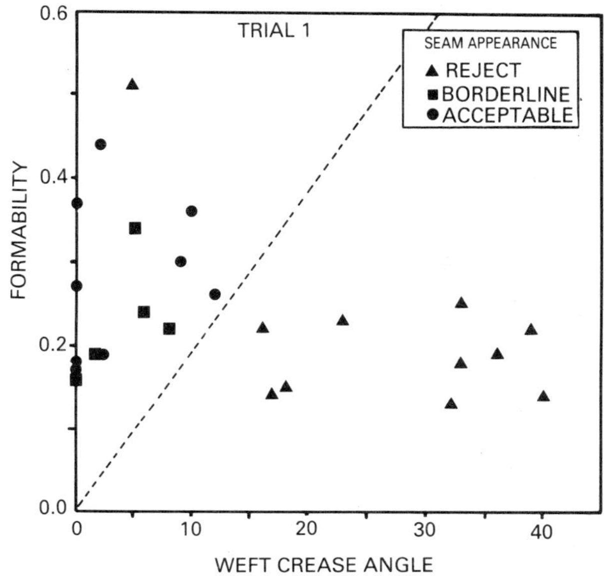

Fig. 4 — Trial 1 — Warp formability vs crease angle.

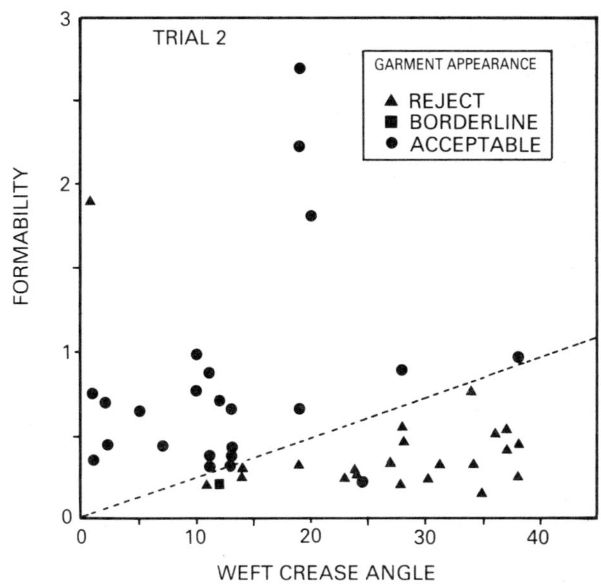

Fig. 5 — Trial 2 — Warp formability vs crease angle.

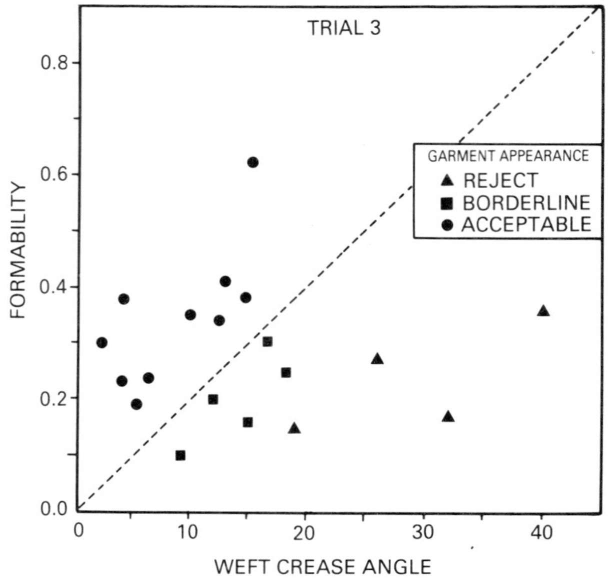

Fig. 6 — Trial 3 — Warp formability vs crease angle.

the final garment. It can be seen that on each graph, there are two clearly defined areas separated by a line going through the origin. Most reject fabrics are located in the area under the curve while acceptable or borderline fabrics are in the area above the curve. The only acceptable fabric that falls in the rejected area of Fig. 5 (A=24; F=0.26) was a fabric accepted for a cheaper, young style line.

4. DISCUSSION

The laboratory trial shows that for each seam a maximum crease angle value of about 15° is acceptable with respect to appearance. This trial was limited to light-weight fabrics in the range 140 g/m^2 to 180 g/m^2. If we look at the behaviour of fabrics in the same weight range in the two industrial trails we get a similar picture. That is, there is a cut-off observed in crease angle above which no garment has a good appearance.

Analysis of the influence of fabric weight on garment appearance suggests that for a seam to appear good, light-weight fabrics require a lower crease angle (or good pressing performance) while medium or heavy-weight fabrics can tolerate a higher crease angle (or border-line pressing performance).

However, a low crease angle does not guarantee good seam appearance. The major contributor, after crease angle, was found to be fabric formability. Across the full range of fabrics used in the industrial trials, there was seen to be a relationship between crease angle and formability which produced a dividing line between acceptable or reject garments.

The graphs on Figs 4, 5 and 6 show a similar trend. Fabrics with a low crease angle can tolerate a lower level of formability and still produce a seam with acceptable appearance, however fabrics with a larger crease angle require a higher level of formability to achieve the same acceptable appearance. This can be explained in the following way.

The appearance of a garment is the combination of a number of factors including sharply pressed seams with low pucker and smooth moulding into three dimensions. It is known by garment makers that there is often a correlation between poor garment appearance and poor crease retention on the lower edges of sleeves and jacket body. Crease angle is related to pressing performance in that the lower the crease angle, the easier it is to press a fabric into a sharp crease. Formability is related to seam pucker. The lower the formability the more likelihood of seam pucker. Fabrics with a low crease angle or efficient pressing performance can tolerate a lower level of formability or worse seam pucker than fabrics with a higher crease angle or less efficient pressing performance.

While crease angle is an indicator of one aspect of pressing performance, a very low crease angle can cause embossing, where an imprint of the seam allowance or underlining becomes visible on the face fabric. One large garment maker has used a very low crease angle fabric for checking the quality of linings, as any texture irregularity would show through the fabric itself.

A constant ratio between crease angle and formability is the dividing line between accepted and rejected garments. The difference in ratios between Figs 4, 5 and 6 suggests that the actual value of the ratio may vary from one manufacturer to another and perhaps also between drastically different styles. Fabrics with a ratio above the dividing line are accepted whereas those with a value below are rejected.

5. CONCLUSION

The measurement of crease angles has proved useful in the prediction of pressing performance, which governs an aspect of final garment appearance. There is an upper limit for the crease angle of light-weight fabrics above which the appearance of garments is not satisfactory. This upper limit was found to be about 15° for the three trials reported in this paper.

Fabric formability is an additional parameter that can be used to identify fabrics causing poor garment appearance. The ratio between warp formability and crease angle provides the threshold for acceptability. Fabrics with a ratio above this threshold are accepted, whilst those with a ratio between this threshold are rejected. This threshold of acceptability may vary from one company to another.

It is not possible to produce a universal threshold for a particular fabric. This is because of other variables in the manufacturing process such as tailoring methods, garment style, pressing machinery, sewing machinery or operator experience, and the requirement for garment quality. Single threshold values may be useful within a particular company, for a specific garment design. However, understanding the individual fabric properties gives wider scope for production planning and control.

Data on fabric crease angle is being collected and assessed at two participating textile companies with a view to including the crease angle test as an additional fabric objective measurement for predicting garment appearance.

ACKNOWLEDGEMENTS

The authors would like to acknowledge the cooperation and support of the management of both VBC, Italy and GFT, Italy, the assistance of Mr G. Grupallo and Mr P. G. Minazio, of the International Wool Secretariat, Italy, in liaising between authors and the help of Mrs I. Sarlej, of CSIRO, with the laboratory trial.

REFERENCES

[1] J. Lindberg, L. Waesterberg and R. Svenson, Wool Fabrics as Garment Construction Materials, *J. Text. Inst.*, 1960, **51**, T1475–1493.

[2] T. J. Mahar, R. C. Dhingra and R. Postle, (1982), The Investigation and Objective Measurement of Fabric Mechanical and Physical Properties Relevant to Tailoring, In: *Objective specification of fabric quality, mechanical properties and performance* (Eds S. Kawabata, R. Postle and M. Niwa), Textile Machinery Society of Japan, Osaka, p. 301.

[3] R. Postle and R. C. Dhingra, 'Tailorability of Woven and Knitted Outerwear Fabrics', Proc. 6th Int. Wool Text. Research Conf., Pretoria, South Africa, 1980.

[4] A. F. Roczniok, U. Biglia and N. G. Ly, How Specific Tailoring Operations are affected by Fabric Mechanical Properties, In: Proc. 8th International Wool Textile Research Conference, Christchurch, New Zealand.

[5] N. G. Ly, D. H. Tester, P. Buckenham, A. F. Rocznniok, M. Brothers, F. Scaysbrook and S. de Jong, Simple Instruments for Quality Control in a Tailoring Company, IWTO Tech. Cttee Meeting, Paris, December 1988, Report No. 11; also, submitted to *Textile Res. J.*

QUESTIONS

MUNDEN: Very interesting piece of work. You measure the recovery angle after 30 minutes and after 24 hours. Which are the figures that you have used? Some years ago we at Leeds attempted to measure crease angle with difficulty because of the weight on the woven fabric. We took the yarn out separately and measure the angle of the individual yarns, that way we got more reliable results but it was time-consuming.

BIGLIA: I have used the recovery angle value after 24 hours. We leave only 1 cm of fabric with the weight on top, this is a better measure in fabrics and not withdrawing the thread due to difficulty in manipulation.

MUNDEN: There was difficulty to ensure that we left 1 cm of fabric because any difference can affect the crease angle.

BIGLIA: We cut the fabric accurately by using a guillotine and a ruler 1 cm apart, so it is easy to get 1 cm. We use 8 measurements all together; 4 and 4 in each direction and we average the results out. What we found is that this angle is quite stable and quite firm, you can move the fabric, it is something inherent to the fabric and the pressing cycle of course, if you alter the cycle you vary the condition but once that you have all the conditions the same, this angle is something that relates to the fabric.

POSTLE: I would like to make a comment in order to relate a few points. This is an interesting paper in the context of a few previous papers. The introduction to crease angle is really related to the recovery behaviour of the fabric. In earlier papers there was some discussion to whether hysteresis was important or not in the whole question of objective measurement. This came yeasterday and also this morning. In a sense you are talking about this point in a different way which I find interesting. Hysteresis is one aspect of non-recovery of the fabric and crease recovery angle is another. In a sense you are including the inelastic properties of the fabric with the formability by including the crease recovery angle. In a way, in my mind, this answers some questions that have been asked before.

BIGLIA: This is a sort of hysteresis. This is a behaviour to steam. It works like hysteresis but it is reduced by steam and by pressing. We can have a fabric with exactly or very similar mechanical properties, but with completely different crease recovery on pressing behaviour. It works like hysteresis, because it stabilizes fabric in a deformed position, but it has nothing to do with mechanical properties.

POSTLE: Yes but nevertheless having had a perfectly elastic fabric under pressing you would get crease recovery.

BIGLIA: No we act on the fabric outside the mechannical properties; we modify the fabric. I do not know what it is, I cannot find anything about it, but I can produce the fabric with the snake chart and a good pressing performance.

Session 4: Robotics and automation in garment manufacture

CHAIR: Professor Paul Taylor
Robotics Research Unit, Department of Electronic Engineering, University of Hull, UK

INTRODUCTION

It is my privilege and pleasure to contribute to this, the last session of this conference, which is on 'Robotics and automation in garment manufacture'. There are four papers in this session, three papers deal with knitted materials and one with woven. We now move from objective measurement to work that people are doing primarily to devise automation systems to tackle handling problems within the industry.

Introduction to Dr I. Gibson

The first speaker is Dr I. Gibson, from the University of Hull, where he completed his Ph.D. research on the automation of shoe sole assembly in a collaboration with SATRA project, he then worked in industry for two years as systems engineer on visual inspection systems. Since April 1989 he has been a senior research assistant in a research project we have with Corah plc and funded by the SERC on automated garment assembly. So he is going to tell us about one aspect of this work associated with the fabric top taping handling operation. The paper has the title, 'Robotics applied to top taping of men's Y-front briefs' and is co-authored by A. J. Wilkinson.

Robotics applied to the top taping of men's Y-front briefs

I. Gibson, P. M. Taylor and **A. J. Wilkinson**
Robotics Research Unit, Department of Electronic Engineering, Hull University, UK

ABSTRACT

This paper describes in detail one aspect of the *fully automated garment assembly* system being developed at Hull. Two garment styles have been chosen for this assembly line, men's Y-front and ladies' briefs. Whereas most of the sewing of these garments is in a straight line, the *Y-taping operation* (i.e. joining the front piece to the back-gusset combination) must be sewn as a complex curve. The reasons for this are discussed.

A UMI RTX robot has been used for this cell and a description of the practical considerations for *large fabric panel handling and sewing* is included along with the software developed for this specialized task. Results taken from the cell to date are included in this paper and the conclusions discuss the impact of these results and the further considerations required to *integrate* this operation with the other *assembly cells*.

1. INTRODUCTION

The Robotics Research Unit at Hull University is performing a number of research activities in connection with garment manufacture. These projects have evolved over a ten year period, from initial investigations into ply separation [1] and fabric properties [2], through individual robotic sewing tasks to the present-day goal of a fully automated garment assembly system.

With the garment industry under increasing threat overseas manufacturers, it is necessary for engineering research activities to look at ways of increasing productivity without affecting quality. One way of achieving this is for the industry to increase the level of automation. Fully robotized garment assembly, with the minimum of manual intervention, is the ultimate end in this approach. The Hull garment assembly system is designed to mimic the manual principle of the assembly line. In this, each task is performed by one or more robots, passing subassemblies on to subsequent stages. Robot gripping mechanisms are kept simple, optimizing their capacity for reconfiguration. This is opposed to some other systems that use one robot to perform multiple tasks, or those with highly complex gripping or sensing mechanism [3,4].

2. CORAH GARMENT ASSEMBLY SYSTEM

The Hull garment assembly system (jointly funded by ACME and Corah plc) is a prototype assembly line for the robotic manufacture of a complete garment. The garments chosen for the initial study were one style each of both ladies' and men's underwear. The ladies' garment is a single rear concealed gusset construction, the men's style is of a rear concealed, Y-front design. Consideration was made at the start for differences in fabric properties and therefore the variations in their handling. These types of garment are most commonly made from single and double knitted cotton based materials.

The assembly system is designed to construct the garment all the way from cut stacks of individual garment pieces, through to the finished item (Fig. 1). This system is being developed in three phases, each running concurrently, to be integrated at a later date. The first phase of this system involves the garment construction in a flat form by sewing the gusset seams to make a 2-dimensional net of the garment. These operations can be seen to vary considerably between the men's and ladies' styles. Once this has been completed, the subassembly is then put through what is termed the 3-dimensional assembly phase. This involves leg taping or elastication, and folding the front and back to mate together, enabling sewing of the side seams and the waistband. The operations for this phase are the same regardless of whether the style is men's or ladies' but perhaps requiring different machine attachments. The third phase of this system involves a production line supervisory system. This will allow for software and hardware reconfiguration between styles and sizes.

Most of the sewing operations for this garment are performed along straight lines. The side seams and the gussets are made up in the main out of straight lines. For those operations, like the leg curve, that are not like this can be made straight using a flexible curve straightener [5]. This device will straighten out any concave curve like the leg or the back of the ladies' brief. The sewing operations can then be performed using linear feeding mechansims, with a feed rate linked to the speed of the sewing machines. This greatly simplifies automated sewing, requiring at the most only small, localized adjustments of the sewing paths using commercially available attachments.

Unfortunately, not all operations can be performed in this manner. One operation that cannot be sewn using a curve straightener is the top taping of the men's Y-front. There are a number of reasons why this is the case:

(1) The construction of the top taping operation involves multiple thicknesses of fabric. Curve straightening would result in an incorrect relationship between the

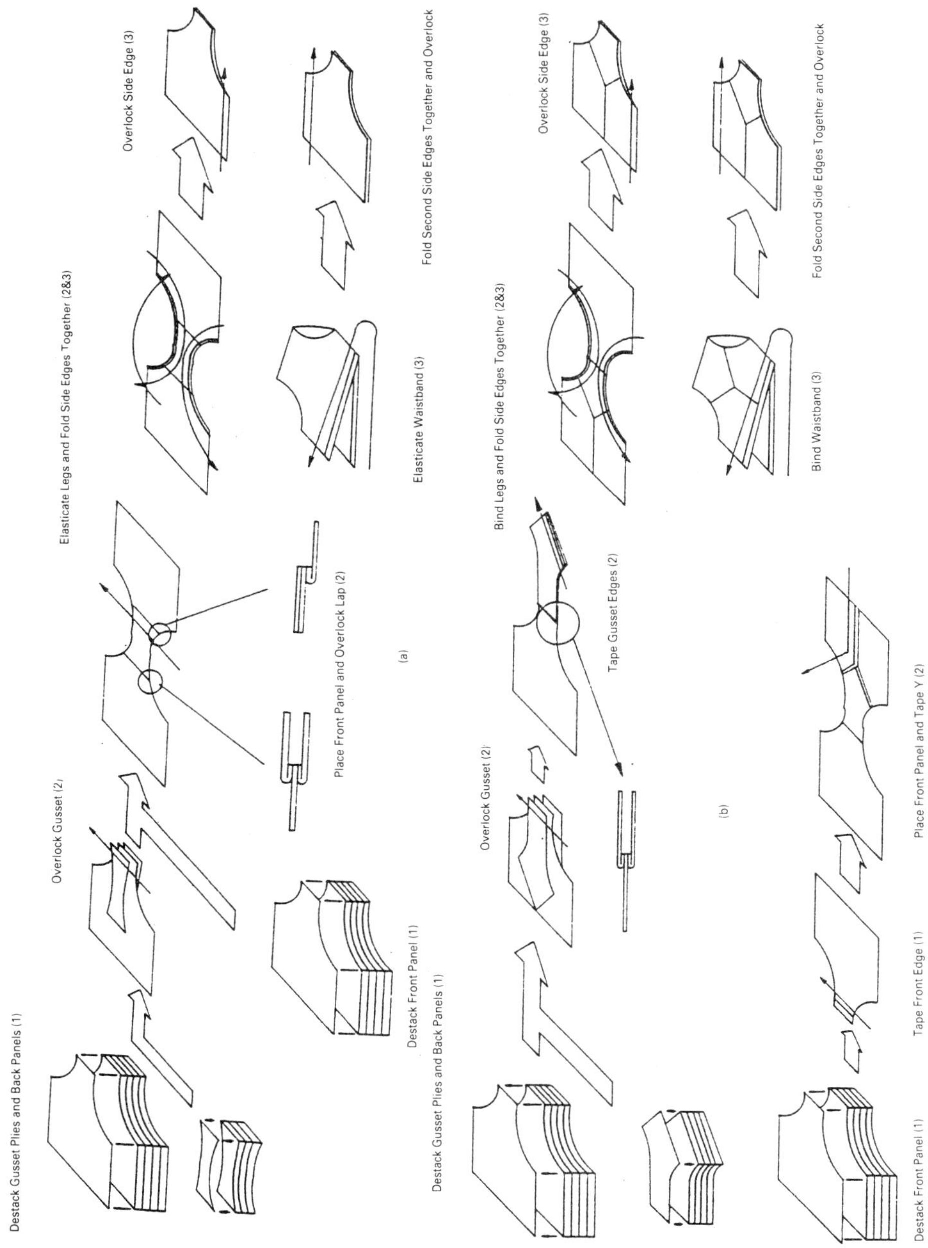

Fig. 1 — Construction of (a) ladies' briefs and (b) men's Y-fronts.

gusset pieces and the front piece since they would move in correlation with each other.

(2) Curve straightening works because there is a defined curve displayed by the cut of the fabric. There is very little fabric on the inside of this curve. The top taping operation sews down the middle of the fabric on the front piece. There would be uneven tensions as a result of curve straightening that would result in pucker during sewing. The same problem would result to a lesser degree from the straightening out of convex curves.

(3) The accepted design for the top taping is for as gentle a curve as possible to be described, within the constraints of the garment size. Such a curve is much more gentle than that described by the component fabric pieces. If the fabric were sewn using a curve straightener then there would be too sharp a corner, resulting in unsightly wrinkles on the inside edge.

It follows therefore that the robotic device for the Y-taping operation must mimic the manual process, steering the garment through the sewing machine.

The actual curved shape of the Y-front serves little and is mostly cosmetic. Since sewing along a straight line is much more simple to automate, requiring only linear feeders, a variation in the design of this garment is proposed along the lines of design for automation. The 'lamda' (λ) front design (Fig. 2) represents a way of constructing the garment using only straight lines. This design would aid automation without

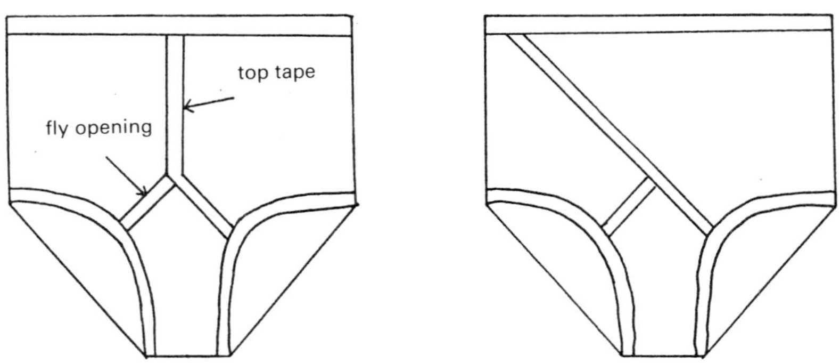

Fig. 2 — Y-front and λ-front garment designs.

affecting the functionality of the garment. However, it is accepted that the concept of curved sewing is valuable in the assembly of other types of garment and cannot be dispensed with in all circumstances.

3. ROBOTIC CONSIDERATIONS

The robot used for the top taping cell was a UM1 RTX robot. This low-cost device is the basis of all the fabric handling operations in the garment assembly system. Driven from a PC under control of software written in Pascal eases the difficulty in construction of new robotic tasks. Similarly, being a robot of light construction and with the adoption within the Robots Research Unit of a universal gripper mount, mechanical reconfiguration of individual workcells is also simplified. The RTX is portable, allowing quick rearrangement of the cell topology and with the departmental robot control software, small test programs can be written very quickly.

The RTX robot was used in conjunction with a Rimoldi pillar sewing machine with a top taping attachment. This arrangement allows for a significant amount of access for a robot mounted gripper on both sides of the needle. This sewing machine was equipped with a Quick Rotan d.c. servo motor and controller which provides a very quick response and accurate digital control of the sewing speed. It is also possible to access this type of controller so that the same PC that is connected to the robot, is also capable of controlling and monitoring the speed of the sewing machine.

It can be seen from the design of the Y-front that the sewing action required is a complex curve. This curve is made up from two straight lines, connected by the arc of a circle. Th e length of the lines, the angle of the arc, and the centre of rotation for the arc all vary with respect to size of garment. It should therefore be necessary for any automated top taping system to allow for reprogramming of these three parameters. Because the centre of rotation varies, it is not possible to build one gripper to account for all sizes of garment using the existing robot control software. This is because there was no facility to control rotational motion using a centre that was not that of the gripper mount position. Because of this, and because this facility is of use to other applications, a tool transform in the XY plane was considered a useful addition to the standard software library. A geometric solution of such a transform can be seen in Appendix A.

4. SEWING CONSTRAINTS

After studying the manual sewing operation for the top taping operation, it was quickly decided that a number of changes were necessary to automate the system.

One manual technique is to sew the first straight up to the start of the curve on the front piece only. The operation is halted at this point, with presser foot lifted, to assemble the remainder of the garment before continuation of the sewing process. To maintain a steady flow of garments on an assembly line basis it was considered that the garment should be assembled prior to sewing. This is mainly because the gripper used for sewing cannot be the same as that for constructing the garment. This would mean that the first gripper would have to be removed and a second inserted, thus wasting time. Similarly, there are difficulties in physically placing two robots in close proximity in such a way that one robot can assemble the remaining pieces after a first has performed sewing of the first straight.

The above approach allows for one continuous sewing operation. A manual operator would have difficulty in performing this because of problems in aligning the garment for a reputable curve. A robot, however, can only move along a programmed path. Even the low cost, low accuracy RTX exhibits a measurable repeatability superior (normally $\pm$ 1 mm) to that of the annual worker.

Similarly, the above consideration means that the presser foot is not raised at any point during the sewing operation. It is only raised at the beginning and end of the cycle. This is advantageous to the robot since the fabric will only move in conjunction with the sewing machine and there is no danger of varying length stitches through pulling of the fabric whilst the foot is raised.

Because there is no presser foot raised throughout the operation then there is a danger of the tips of the arrow-shaped gusset pieces folding prior to feeding under the foot. Similarly, when the sewing machine changes from putting tape on to a single fabric ply to multiple thickness there is a tendency for bad stitching to occur at the interface. The robotic top taping system therefore starts at the trailing end of the garment, sewing tape on to the three thicknesses of material first since this can be moved into proximity of the needle with the presser foot raised. The foot therefore comes down on this multi-thickness, avoiding any feeding problems. Going down from sewing tape on three plies to just a single ply does not have such an effect on the quality of stitching.

Handling of the garment is a little more difficult in the automated process since the garment itself is larger at this stage of the assembly. The reason for this lies in the gusset construction. It would be extremely difficult to perform the inverted gusset sewing after the front is fully asembled. The rear concealed gusset seam is therefore sewn first, making this operation far simpler and far outweighing the deterimental effect on the top taping operation.

The gripper used in the top taping system controls the fabric on both sides of the needle. This was necessary to control the direction of the fabric and can be seen to mimic the operation of the manual process. Close control is important in both direction and speed so that a garment design that took the shape of the sewing path was reliably constructed.

5. RESULTS

Top taping was performed on a Rimoldi 'system 64' pillar sewing machine with its original controller replaced by a Quick d.c. servo motor/digital controller combination. The only attachment required to prove the technique was a $1\frac{1}{8}$-inch tape feeder and tensioner. The tape was top fed and oversewn with a standard chain stitch using two sets of needles. A number of successful runs, using the RTX to feed the garment through the machine were performed. There was a measured repeatability along the entire curve of better than ± 5 mm deviation from the curve which is the required accuracy from manufacturers of this style of garment.

The fabric used was a double jersey, cotton-based knitted material as commonly used for underwear. Three components were sewn for each garment, two arrow-shaped gusset pieces placed on a front piece. The gussets were sewn together using binding tape along one straight edge of the arrow. Because the experiments were initially to look at the feasibility of top taping using a robot, the gusset pieces did not have the back piece added. It is not considered that this will affect the performance of the sewing but may require an augmented gripper design with an attachment for taking up the waste material prior to sewing.

Some pucker at the point of the arrow was visible for a few of the garments sewn. This is primarily due to the gripper not being close enough to the sewing head at this point and so there can be a tendency for the material to drift at this point. The gap in the gripper is quite wide in this area to allow for the turning of the curve (Fig. 3). The presser foot is wider than it needs to be and a smaller gap in the gripper, and therefore tighter control can be obtained if the presser foot were cut down.

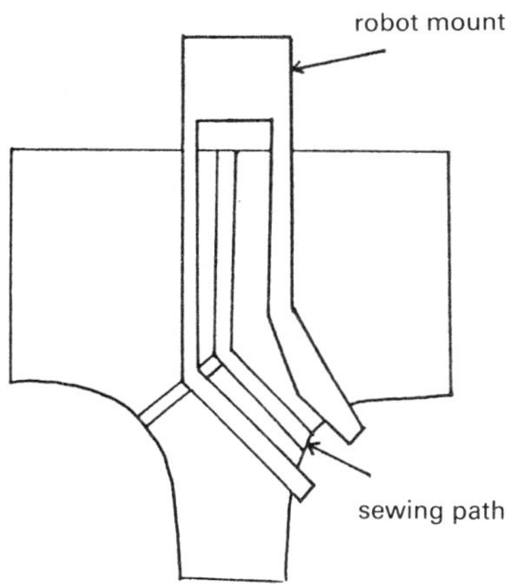

Fig. 3 — Gripper design.

The gripper is a rigid structure of a very simple design. The gap in the gripper can only handle a limited range of sizes and tests were performed that showed curves can be sewn over a range of included angles from 120°–150°. This will cover a range of sizes roughly correspondingly to medium through to extra large.

Placement of the fabric components is critical for a good finished item but it does not affect the trajectory of the curve. If the garment is positioned, prior to sewing, prior to sewing, to within ± 3 mm then the completed item with top tape attached will pass quality control requirements.

Sewing speed for this operation was quite slow (approx. 650 stitches per minute) in comparison to the manual method. This was necessary to match the maximum robot speed on the curve of 26 mm/s. This gives, for a typical large garment, sewing time of 8 s. Combining this with around 2 s for feeding up to the needle and 4 s return time gives a resulting cycle time of less than 20 seconds. A further robotic device would be required to place the pieces and another for removal from the top taper which could also be used for the subsequent garment assembly operation.

6. CONCLUSION

It has been shown that a robot can be used to perform curved sewing operations. Although it has been mentioned that a great many of the functional garment sewing assembly operations can be performed using straight line sewing, curved sewing is performed often enough, and over a wide range of garments to make the results of this experiment significant.

Of equal importance is the fact that a relatively low cost and unsophisticated robot such as the UMI RTX is capable of this type of sewing. The off-axis tool transform is a useful addition to the extensive software library available for this machine. Such a facility has wide-ranging uses outside the subject of garment assembly.

The accuracy of the top taping operation matches up to the requirements for this type of garment. The sewing technique requires little further development but the alignment of the garment pieces requires further thought. Although alignment has been demonstrated [6], further work must be done to produce a quicker time for this operation.

It should be emphasized that this robotic work cell is to be part of a full garment assembly system. There are a number of operations in this system that take as much time, if not more, to complete their cycle. Therefore a slow sewing speed for a relatively short overall path length (compared for example with the leg taping operation) does not have a significant effect on the overall system cycle time. Further, there is potential for using this system with lower cost, slower sewing machines than is traditionally used in the industry.

REFERENCES

[1] Kemp, D. R., Taylor, P. M. and Taylor, G. E., 'Adaptive sensory gripper for fabric handling', Proc. 4th IASTED Symposium on Robotics and Automation, Amsterdam, Holland, 1984.

[2] Taylor, P. M. and Koudis, S. G., 'Automated handling of fabrics', *Science Progress*, Oxford, **71** pp.351–363.

[3] Aisaka, N. 'Technical development in apparel industry', *JTN* (Shirley Institute), **390**, pp. 47–51, 1987.

[4] Gershon, D. and Porat, I., 'Vision servo control of a robotic sewing system', IEEE Proc. of Int. Conf. on Robotics, 1988.

[5] Palmer, G. S., Wilkinson, A. J. and Taylor, P. M., 'The design of an integrated robotic garment assembly cell', Proc. 6th National Conf. on Production Research, Glasgow, Sept. 1990.

[6] Taylor, P. M. and Bowden, P., 'The use for multiple, low cost vision sensors in fabric pick and place tasks', Proc. IFAC Symp. on Low Cost Automation, Valencia, Spain, 1986.

APPENDIX A. ROTATIONAL TRANSFORM

The RTX robot is supplied complete with a communications handler that allows the user to determine the status of the robot. There is, however, limited software support apart from this. Over the years of use, individuals within the Robots Research Unit have developed a number of software tools to make robot programming a much simpler task. Utilities have been developed in a specialized form using the basic status information to allow generation of point to point and joint interpolated motion in a variety of modes. To minimize computing time a general matrix form was avoided and a time slice technique was adopted. This technique is simplified by the fact that the two main arm links are the same length. Even so, it is not possible to generate joint interpolated motion in real time. The points along the path are calculated and then placed in an array which is then assessed at the time when the robot is required to move.

One function that was not available in the software library was an off-axis rotation about a point. This is commonly known as a tool transform. For movement in a plane, this can be realized in the following way, consider Fig. 4.

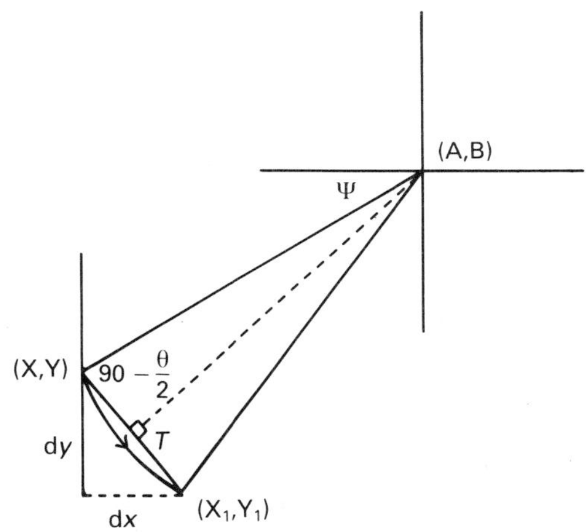

Fig. 4 — Geometric interpretation of tool movement. (A,B), centre of rotation; (X,Y), initial position; (X1,Y1), final position; T, length of chord; Ψ, initial angle offset; θ, desired rotation; dx, dy, offsets between points.

If point (A,B) is the centre of rotation and (X,Y) is the initial position and (X1,Y1) the destination of the gripper then using the sine rule:

$$\frac{r}{\sin(90-\theta/2)} = \frac{T}{\sin\theta}$$

gives

$$T = \frac{r\sin\theta}{\cos(\theta/2)}$$

This is the linear offset or the distance through which the gripper must move for an angular offset θ from the centre of rotation.

To calculate the offsets dx, dy, the angle ω must first be known. This is done using $\tan^{-1}$ after imposing a coordinate system orthogonal to the robot frame of reference but with its origin at the point of rotation. For the example shown the value for dx, dy can be calculated from the angle ϕ

$$\Phi = 180 - \psi - \frac{180 - \theta}{2}$$

$$dy = T \sin \phi$$

$$dx = T \cos \phi$$

The value for ψ is calculated for different quadrants (Fig. 5). It is the angle subtended at the centre of rotation for positive θ and positive slope, or negative θ and negative slope. ψ is the angle at the circumference for the other instances of θ and slope. The slope is calculated from a test of distance from the X,Y vertices:

If $T > 2^*$(distance to nearest vertex in direction of θ) then slope changes sign.

Similarly, the direction of dx, dy can be calculated in this manner.

This solution gives a position for the gripper but not the orientation. It would be advantageous for the gripper to be pointing in the same attitude after rotation as it was before rotation. An adjustment to the yaw position must also be made. Since the elbow and yaw joints are linked in the RTX, then there is a parallel yaw value for any elbow position. The adjustment to be made is therefore a rotation in the yaw of an angle $-\theta$. A gripper pointing towards the centre of rotation prior to movement, will also be pointing towards this point after rotation.

The above gives the point to point movement but it does not necessarily describe the arc and is performed at full velocity. Joint interpolated motion can be achieved by selecting the overall rotational value θ. By splitting up the angle θ into small, equally spaced components, the size of which is linked to the sample rate of the RTX, control of the interpolated velocity is obtained. A further adaptation of the existing software library allows for the creation of complex curve tables without stopping between the simplex values. This is necessary to achieve a smooth sewing action, starting at the beginning of one straight and ending at the end of the other.

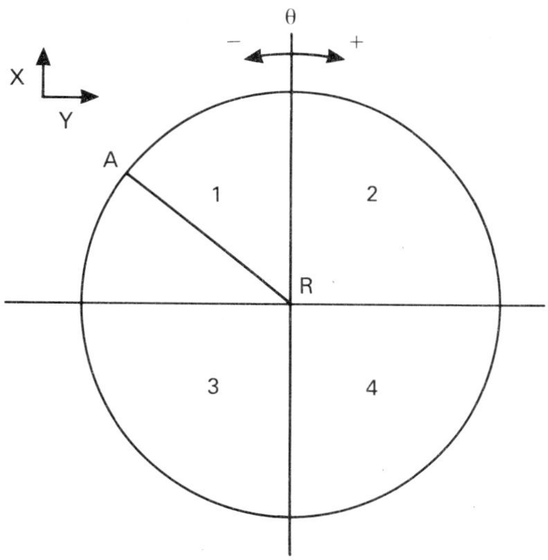

Fig. 5 — Quadrant layout for tool transform. A, actual gripper position; R, centre of rotation.

Quadrant 1:
$\theta+$ if $dy > 2|(Ry - Ay)|$ then $dx = -dx$
$\theta-$ $dx = -dx$
if $dx < 2|(Rx - Ax)|$ then $dy = -dy$

Quadrant 2:
$\theta+$ $dx = -dx$
if $dx > 2|(Rx - Ax)|$ then $dy = -dy$
$\theta-$ $dy = -dy$
if $dy > 2|(Ry - Ay)|$ then $dx = -dx$

Quadrant 3:
$\theta+$ if $dx < 2|(Rx - Ax)|$ then $dy = -dy$
$\theta-$ if $dy < 2|(Ry - Ay)|$ then $dx = -dx$

Quadrant 4:
$\theta+$ $dy = -dy$
if $dy < 2|(Ry - Ay)|$ then $dx = -dx$
$\theta-$ if $dx > 2|(Rx - Ax)|$ then $dy = -dy$

QUESTIONS

SARHADI: How did you devise the alorithm to synchronize numerically the rate of feed with the sewing machine speed.

GIBSON: Its basically feedback from the shaft encoder of the sewing machine. We work out how fast the sewing machine is working in mm per second and we then feed that back to the robot, we can then calculate how fast we need to feed the fabric through, we may have about ±5% deviation, but due to the elasticity of the material we find that this is adequate. What we tend to do is not to adjust the speed of the robots in real time but to keep monitoring the sewing machine cycle. If the sewing machine goes out of a preset error bound it is stopped. Up to now we have not experienced problems in this area.

SARHADI: Do you know how much time the manual operation takes? And how does this compare with your system?

GIBSON: We estimate something like 40 seconds. We have conducted simulations with experimental results, and we think that the overall time may be about 80 seconds, which we are quite capable to improve, particularly with the external controller that we have developed for the RTX robot.

QUESTION: The fabrics you use are perhaps slack. How reliably can you select the top layer? How do you overcome a problem of say double plies, etc.?

GIBSON: This is based on eariler work at Hull; we are talking very close to 100% reliability, particularly since we use sensors to detect double plies. So you can retract and try again or reject. If we fail after a number of attempts then we discard the fabric.

TAYLOR: Just a comment: at this moment this is the most reliable part of the system and it is based on the air-jet finger ply separation device that we developed on a robot (autotex).

SLATER: I presume that some sort of mechanism will stop the entire system if it is jammed. You showed a curve that demonstrates the error on control. What sort of scatter do you get?

GIBSON: Less than a size difference, there was no deviation greater than ± 5 mm in the entire sample.

STYLIOS: What is the density of fabrics used?

GIBSON: No idea; we cope with whatever we get from Corah plc. If you would like us to try particular densities of fabric we would be more than happy to try them. At the moment we try to set the system up then we try different fabrics.

STYLIOS: Are you planning to do any work about faults associated with needle, thread or fabric?

GIBSON: We do not perform any experimental approach in this area. Our main objective is to get the system running. This will be a prototype system, a test rig to enable us to look at these particular aspects, based on the work that people like yourself or others are doing.

SLATER: I presume that you can try to use fabrics with different elasticity.

GIBSON: We can definitely try to see how the system will perform. A part to why we are here is to discuss our approach and to get feedback for developing the system further.

THEVARATNAM: In concluding you have mentioned about the variation of ±5 mm which is effectively 1 cm variation. This appears satisfactory for knitted fabrics; if you use let's say shirts, however, 1 cm is quite a lot.

GIBSON: Yes, I could not agree more, we are looking at knitted fabric and assembly of Y-fronts, there are problems associated with knitted fabrics that you do not have with woven fabrics, so we have to have a different approach. If you require a system within the capabilities of this particular one for other materials or designs then we can isolate the problem and look at ways to solve it.

SLATER: Have you tried to use woven fabrics, because as they do not have elasticity the percentage accuracy may be much higher.

GIBSON: I agree, it will be a good experiment actually; wovens have no or very little elasticity and we can see how well the system is performing. We have tried with paper when investigating the stitch density, of course without sewing thread.

QUESTION: You have said that you can speed the system up and also reduce the payback period, I am interested in the basis on which you have made that statement.

GIBSON: This is a part of a complete system, if the system is coordinated most of the operations perform within a set band of time; nothing is idle within a particular length of time, then you have an optimism system. One way of improving the performance of the one robot that you are using throughout, is by increasing the performance of the robot using the external controller that we have developed. Hence you are increasing your throughput because every operation will be speeded up by the same amount.

This particular operation takes about 30 seconds because it is not the slowest operation in the entire cycle, our slowest operation takes about 80 seconds. We are looking into speeding that up of course. We have done some rudimentary costings on the system on how long would take to pay back, based on figures that we had from Corah plc and a three-year payback was feasible and competitive, based on an 80-seconds figure for every cycle. As I say, if you speed this up, you reduce your payback period. If we replace the RTXs with something that is faster, so that the sewing operation becomes faster, it will take less time to pay back.

QUESTION: Is there no operator involved in this system?

GIBSON: Yes, this is what we are aiming for. Obviously there will be a maintenance person but what we are considering is the question of what if we remove one robot from the system, does this bring the entire system down to a halt? By removing a robot we can replace it by a manual operator who can perform the same operations as well as coordinate and communicate with the controllers of the entire system.

QUESTION: How flexible is this system designed to be?

GIBSON: It is designed to be flexible and cope with different garment styles as well. I have shown you Y-fronts; virtually the same system will be used to produce ladies' briefs.

QUESTION: What will be the change over time?

GIBSON: That is difficult to say at present, I guess a matter of minutes not hours. This is a prototype system and it is difficult to say.

THEVERATNAM: Do you have any time-scale to complete the work of this project?

GIBSON: We are aiming to produce a garment the end of this year. I am not going to guarantee that you are going to want to wear it though.

THEVARATNAM: We might stop selling Y-front briefs at the end of the year.

Introduction to Professor Trevor Little

Our next speaker is Professor Trevor Little from North Carolina State University. Trevor Little is currently Professor and Director of the Apparel Programme. He was awarded his B.Sc. and Ph.D. in 1974 from the University of Leeds; he is working in the area of automation and material properties, and on objective measurement. He is a member of numerous committees in the USA in particular at the Textile/Clothing Technology Corporation at Raleigh, and is a member of the Editorial Board of the *International Journal of Clothing Science and Technology*. The title of the paper is 'Sewing dynamics: influence of fabric properties and sewing machine settings on fabric feeding', and has as joint authors: B. A. Matthews, C. G. Carrere and D. Vass.

Sewing dynamics: influence of fabric properties and sewing machine settings on fabric feeding

T. J. Little, B. A. Matthews, D. Vass and **C. G. Carrere**
College of Textiles, North Carolina State University, Raleigh, USA

ABSTRACT

Today's apparel manufacturing industry is faced with the production of *smaller lots* in a *faster time* but at the same time increasing the percentage of *first-quality product*. This is a challenge to *automation* since frequent changes of fabrics as well as styles cause extensive set-up times for the equipment. Any system that allows for *consistent machine settings* for a given fabric from lot to lot and workstation to workstation will enhance the *apparel assembly process*.

The influence of fabric properties on sewing performance, especially the *feeding of plies* of textile materials through a sewing machine were studied on a sewing machine equipped with *high dynamic response transducers* on the *presser bar* and *needle bars* of a Pfaff 463 lockstitch sewing machine. The forces experienced by the presser bar were recorded and the displacement of the presser bar was measured simultaneously using a displacement transducer. The paper discusses the correlation of *feeding forces* with *fabric properties* and the effect of changing sewing speed and sewing combination. *Presser foot bounce* is found to be dependent on *speed, sewing combination* and the *properties of the materials* being stitched. Presser bar displacement and *force waveforms* are analysed both in the time and frequency domains to evaluate how material properties can be used to control or adjust machine setting in an *automated apparel environment*.

INTRODUCTION

Automating the assembly of apparel products continues to present many challenges for clothing manufacturers and machinery developers. This is especially true for the production of small lots in a Quick Response business environment when the number of fabric selections and machine set-ups are frequent and complex. Textile fabrics can be chatracterized by measurement of the physical properties and many researchers have used fabric property data in designing clothing assemblies and modifying apparel pattern data to take into account the different drape or garment size resulting from fabric properties. Apparel computer-aided design systems promoting these features are already available commercially [1,2]. Other researchers have used fabric property measurements systems to assist in selection of other components of the final garment, such as fusible interlinings. The resultant fused composites have been characterized and analysed in relation to the properties of both the shell and interlining fabric properties [3]. Additional research has also been conducted to assist in predicting the amount of overfeed or underfeed neccessary to join seams with the correct amount of fullness and/or ease [4–14].

An automated workstation of robotic work cell used to assemble apparel will require that many of the adjustments be preprogrammed or adjusted automatically rather than be made manually. These adjustments might include thread change, needle change, bobbin/looper thread change, thread tensions, presser foot type, presser foot pressure, feed dog movement, top feed movement (if available), machine speed, edge guidance control, parts loading, parts orientation, parts sewing and part stacking.

The result of the complex interactions between fabric/machine/thread should be a seam of high quality that permits the next assembly operation to be performed satisfactorily. This paper studies the process of seam formation and, in particular, how fabric properties influence feeding of fabric on a drop feed sewing machine. Problems that are known to occur include seam puckering, seam gathering, ply slippage, incorrect seam allowances, fabric fraying, etc.

In attempting to automate the sewing process the properties that control the dynamics of fabric feeding must be identified so that appropriate machine control functions can be developed. The work of Johnson [15] analysed the feeding on Singer machines and showed that stitches per inch were dependent on both the presser foot pressure and the machine speed. It was also observed that there was a difference between types of machine for the fabrics used in the experimentation. The general tendencies were that the stitches per inch decreased with machine speed and were greater for high foot pressures. The differences between the nominal stitch settings and the stitches per inch in the seam were attributed to differential friction effects and the occurrence of presser foot bounce. Other researchers have related fabric mechanical properties to tailorability and sewability [16–33].

Objective of this work

The primary objective of this present work is to understand the relationship between fabric mechanical properties and the sewing process. Although other work has studied the sewability aspects in relation to mechanical properties this paper deals only with the process of fabric feeding with a drop feed sewing machine. For this work it was necessary to place measuring transducers on the sewing machine so that the forces involved in fabric feeding could be quantified.

Experimental

A Pfaff model 463 drop feed lockstitch machine was equipped with transducers on the presser bar. A Kistler 9001 load washer was inserted in the presser bar just above the presser foot mounting and a Kman proximity transducer was mounted directly above the extended presser bar on top of the sewing head. The signals from these transducers were recorded on a digital oscilloscope. This experimental set-up facilitated the recording of feeding force waveforms and the displacement of the presser bar. The data acquisition system is shown schematically in Fig. 1.

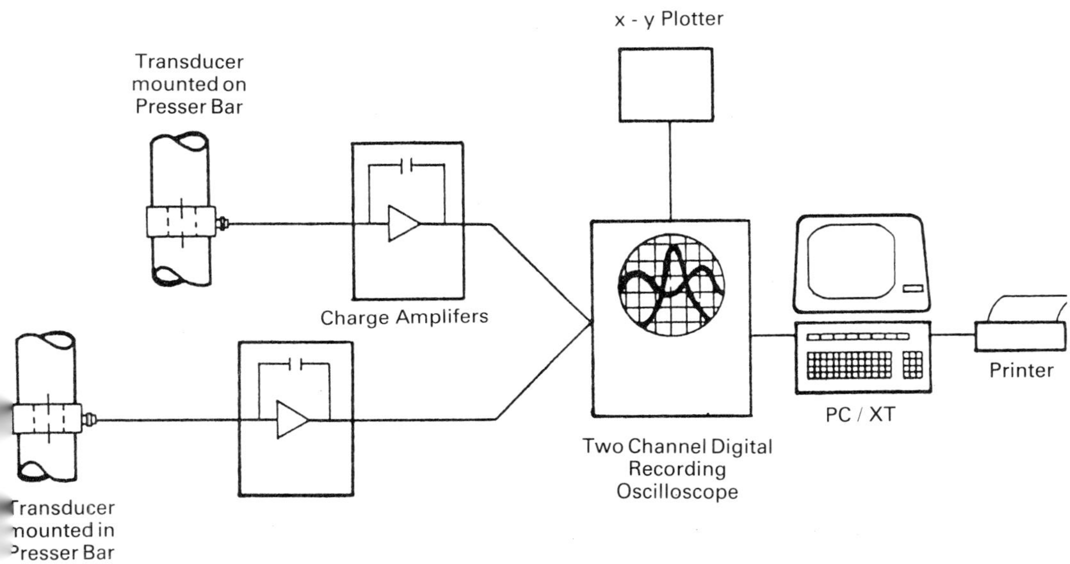

Fig. 1 — Component configuration of the NCSU sewing dynamometer.

Materials and attachments

Presser feet

Several different models of presser feet were used in the trials to investigate how the behaviour of the presser foot assembly might influence fabric feeding characteristics. The different presser feet were

(a) double-compensating presser foot
(b) hinged-spring presser foot
(c) hinged long-base presser foot

The corresponding throatplates and feed dogs for the presser feet were fitted for the experimental trials characterizing the machine response at a variety of sewing speeds.

Fabrics

Four fabrics were selected with different weights and thicknesses to examine the behaviour of the sewing machine when feeding fabric at different speeds. These fabrics are as in Table 1.

Table 1

	Weight (mg/cm^2)	Thickness (mm)
Fabric 1	51.43	0.98
Fabric 2	25.22	0.43
Fabric 3	21.06	0.43
Fabric 4	23.12	0.64

An additional set of fabrics, as shown in Table 2, was chosen for the experimental trials correlating fabric properties with fabric feeding so as to include a reasonable range of weights (13.45–42.78 mg/cm^2), thicknesses (0.393–1.472 mm) structures and apparel end uses. Each of the fabrics were characterized by the KES system at

Table 2 — Fabric weaves, contents, and description

Fabric No.	Fabric weave	Content	Fabric description
1	Plain	100-cotton	Chambray
2	Ground pick	100-cotton	Corduroy
3	Ground pick	50-cotton/50-poly	Corduroy
4	2×1 LH twill	100-poly	Gaberdine
5	Plain	50-I.T. rayon[a] 50-poly	Shirting
6	Plain	100-I.T. rayon[a]	Shirting
7	Plain	100–800 rayon	Shirting
13	Plain	100-cotton	Shirting
14	2×1 LH twill	100-cotton	Denim
15	2×2 basket weave	100-poly	Suiting
16	Plain	81-poly 12-linen 7-silk	Suiting
17	Plain	65-poly 35-wool	Suiting
18	Herringbone	55-poly 45-wool	Suiting

[a]Denotes Improved Tenacity Rayon.

standard conditions and the warp and weft values are given in Tables 3 and 4, respectively.

THE EFFECT OF SPEED ON FABRIC FEEDING ON A DROP FEED MACHINE

Experimental design

The experimental design was set up as a full factorial with the three sewing combinations, four fabrics, eight levels of machine speed and each factor combination was replicated five times. The machine speeds used for these trials were 163, 338, 619, 1017, 1900, 2600, 3886 and 4600 rpm.

Typical feeding-force waveform

A typical feeding-force waveform sewing two plies of fabric 1 with the double-hinged compensating presser foot is shown in Fig. 2. The legend for Fig. 2 is as follows:

Point 4 is the rise of feed dog
Point 5 is advancing of the feed dog
Point 1 is where the feed dog is descending
Point 2 is where the feed dog has descended below the throatplate
Point T is where the needle is at the top of its stroke
Point P is where the needle is penetrating the fabric
Point B is where the needle is at the bottom of its stroke
Point W is where the needle withdraws from the fabrics

The fabric feeding period of the sewing cycle, i.e. when the feed dog is above the throatplate, appears to have three definite stages at high speeds (4600 rpm). The first stage appears when the feed dog rises and begins to advance the material (Points 4 and 5) showing an initial increase in force. This is followed by a second stage where the force on the presser bar decreases significantly coinciding with the descent of the feed dog (Point 1). This is followed by a third stage showing a large peak force at the end of the feeding period as the feed dog descends below the throatplate. The magnitude of this final peak force is dependent on the number of plies being sewn and the dampening effects of the material.

It will also be noted that the presser bar experiences cyclical forces during the stitch formation period of the sewing cycle.

Results

Figs 3, 4 and 5 show the effect of sewing speed on the forces experienced by the presser foot. These figures correspond to the responses of the three sewing combinations chosen for experimentation and show the range of forces on each presser foot. The variables affecting the transfer of energy to the presser foot are primarily:

(i) the mechanics of the presser foot itself
(ii) the impact speed of the feed dog

Table 3 — KES values — warp direction

Fabric no.,	EM%	LT	WT	RT%	G	*2HG*	*2HG5*	*MIU*	*MMD*	*SMD*	*LC*	*WC*	*RC%*	*B*	*2HB*	*T*	*W*
1	7.830	0.314	6.150	31.301	0.740	1.200	2.300	0.211	0.0240	7.310	0.376	0.189	42.33	0.0430	0.0450	0.553	13.90
2	1.070	0.919	2.400	54.000	1.730	2.450	5.380	0.237	0.0114	5.980	0.419	0.457	51.91	0.2400	0.2500	1.240	32.70
3	1.180	0.954	1.740	53.800	2.320	3.420	9.120	0.252	0.0113	5.560	0.368	0.293	42.29	0.3800	0.4350	1.472	35.58
4	3.253	0.568	5.120	50.400	0.790	1.293	3.455	0.208	0.2300	5.000	0.404	0.408	46.30	0.1195	0.0700	1.207	25.99
5	1.170	0.817	6.090	54.520	0.920	0.960	3.330	0.179	0.0517	9.350	0.417	0.135	53.13	0.0640	0.0610	0.451	16.58
6	1.160	0.810	4.980	49.900	0.870	0.870	3.290	0.177	0.0467	9.060	0.415	0.132	52.23	0.0630	0.0580	0.442	15.58
7	1.030	0.809	2.090	50.300	0.670	0.670	3.510	0.167	0.0338	9.040	0.416	0.133	51.56	0.0750	0.0640	0.433	13.45
13	2.200	0.319	6.280	49.370	1.440	2.220	4.660	0.199	0.0350	6.573	0.297	0.238	43.37	0.07230	0.0820	0.645	19.38
14	5.470	0.579	15.730	30.580	3.170	3.840	7.720	0.172	0.0297	7.560	0.227	0.138	42.05	0.2560	0.5270	0.965	42.78
15	14.740	0.674	21.055	48.200	0.760	1.475	3.000	0.247	0.0223	12.530	0.491	0.149	49.90	0.1970	0.0540	0.839	23.82
16	2.430	0.732	4.450	47.191	0.380	0.600	0.950	0.197	0.0200	12.230	0.363	0.327	48.32	0.1260	0.0860	1.038	25.90
17	3.500	0.651	5.700	51.754	0.600	0.630	1.550	0.153	0.0200	3.380	0.380	0.113	61.94	0.0440	0.0210	0.473	18.30
18	3.470	0.749	6.500	62.308	0.810	0.850	2.100	0.170	0.0800	8.230	0.420	0.091	60.44	0.0630	0.0280	0.393	15.20

Table 4 — KES values — weft direction

Fabric no.,	EM%	LT	WT	RT%	G	*2HG*	*2HG5*	*MIU*	*MMD*	*SMD*	*LC*	*WC*	*RC%*	*B*	*2HB*	*T*	*W*
1	11.070	0.317	8.775	34.758	0.810	1.200	2.500	0.215	0.0290	6.350	0.376	0.189	42.33	0.0210	0.0230	0.553	13.90
2	8.670	0.750	16.300	48.500	1.920	2.230	5.630	0.261	0.0275	38.350	0.419	0.457	51.91	0.0900	0.0800	1.240	32.70
3	7.860	0.848	12.720	49.640	2.580	2.980	9.560	0.268	0.0247	30.000	0.368	0.293	42.29	0.0120	0.1530	1.472	35.58
4	4.367	0.696	6.700	59.800	0.940	0.975	4.265	0.222	0.1900	4.400	0.404	0.408	46.30	0.1125	0.0480	1.207	25.99
5	2.790	0.813	11.090	49.780	0.858	0.880	3.170	0.171	0.0187	5.990	0.417	0.135	53.13	0.0520	0.0430	0.451	16.58
6	2.770	0.798	11.080	44.800	0.880	0.910	3.320	0.167	0.0175	5.980	0.415	0.132	52.23	0.0500	0.0390	0.442	15.58
7	2.700	0.795	5.370	38.200	0.580	0.580	3.840	0.166	0.0189	5.380	0.416	0.133	51.56	0.0620	0.0460	0.433	13.45
13	2.700	0.327	8.820	54.030	1.760	1.980	5.340	0.207	0.0390	5.149	0.297	0.238	43.37	0.0540	0.0600	0.645	19.38
14	6.370	0.531	16.040	28.340	3.250	3.930	7.110	0.184	0.0343	3.058	0.227	0.138	42.05	0.2040	0.4590	0.965	42.78
15	12.860	0.624	23.345	51.200	0.712	1.283	2.880	0.225	0.0277	11.570	0.491	0.149	49.90	0.1870	0.0856	0.839	23.82
16	5.480	0.536	7.350	41.497	0.350	0.580	0.920	0.206	0.0170	9.860	0.363	0.327	48.32	0.0821	0.0520	1.038	25.90
17	3.100	0.632	4.900	60.204	0.525	0.280	1.130	0.176	0.0180	3.060	0.380	0.113	61.94	0.0650	0.0190	0.473	18.30
18	4.170	0.714	7.450	59.060	0.830	0.650	2.000	0.185	0.0590	7.790	0.420	0.091	60.44	0.0540	0.0240	0.393	15.20

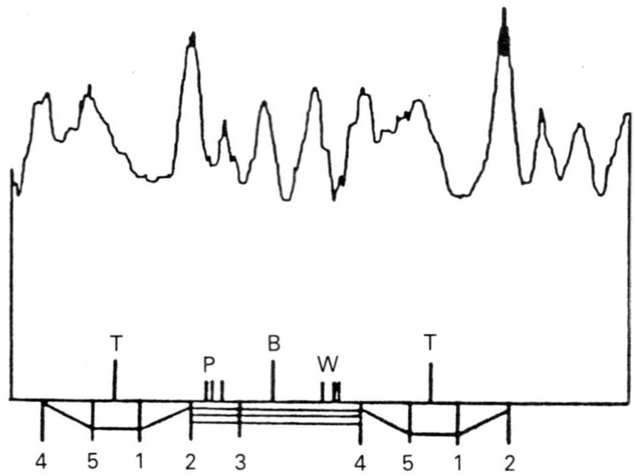

Fig. 2 — Typical feeding force waveform, two plies of fabric.

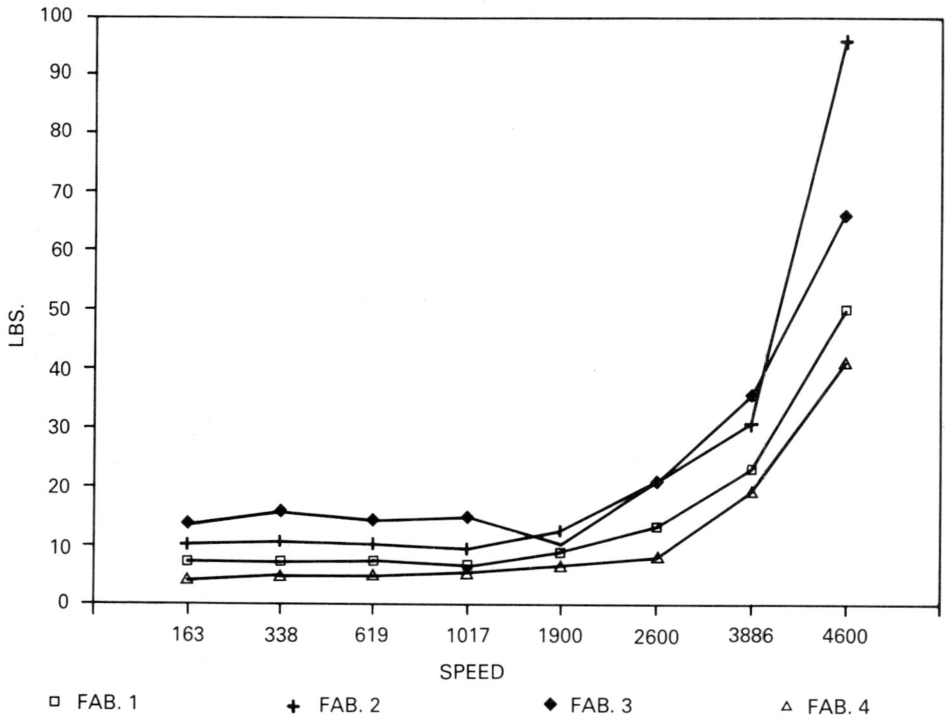

Fig. 3 — Combination 1 — range of forces exerted on the presser bar for eight levels of machine speed. Compensating presser foot used.

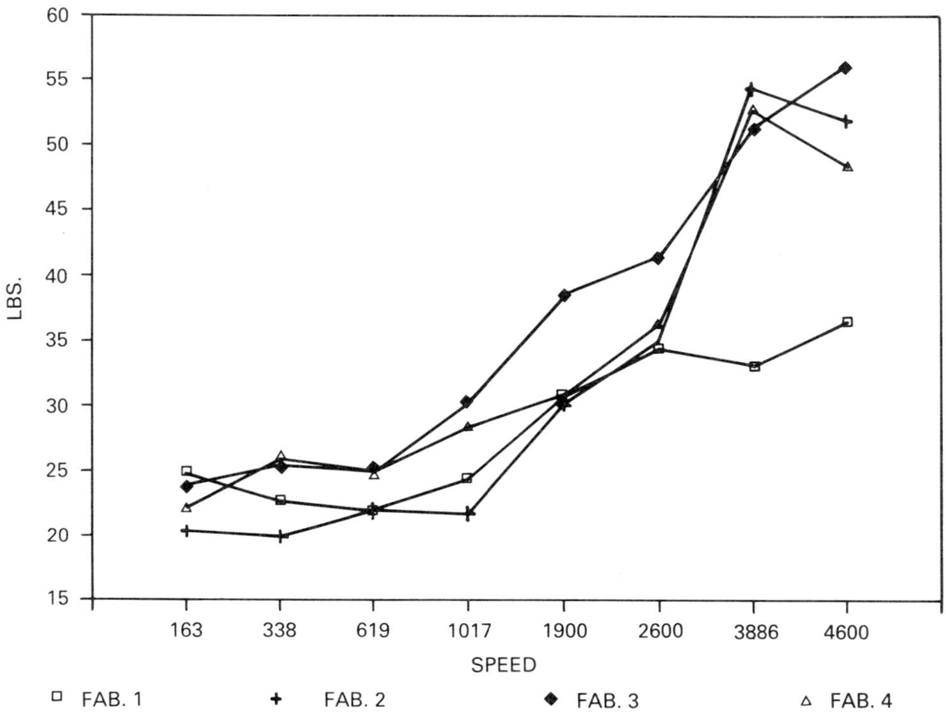

Fig. 4 — Combination 2 — range of forces exerted on the presser bar for eight levels of machine speed. Hinged-spring presser foot used.

(iii) the characteristics of the material being sewn.

Fig. 3 shows that the double-hinged compensating foot sewing combination has a low feeding force below sewing speeds of about 2600 rpm but above 2600 rpm and especially above 3886 rpm the feeding forces increase rapidly. It will also be noted that at the lower speeds there is similar responses for each fabric but at the higher speeds the differences due to fabrics become more evident.

When the sewing combination is changed to the hinged-spring presser foot (Fig. 4), not only are the feeding forces greater but, in addition, they increase with increasing speed especially above 619 rpm. With this sewing combination the differences due to fabric characteristics appear greater at higher speeds.

The third sewing combination (Fig. 5) shows that the feeding forces depend on the fabric types being sewn and only with fabric 1 is there a dependence of feeding force with machine speed.

These result suggest that the fabric feeding forces as experienced by the presser bar are dependent on the sewing combination, the sewing speed and the type of fabric being sewn. It was decided therefore to investigate which material characteristics contributed to the feeding forces so that a knowledge of the critical fabric properties could lead to automated machine changes and adjustments.

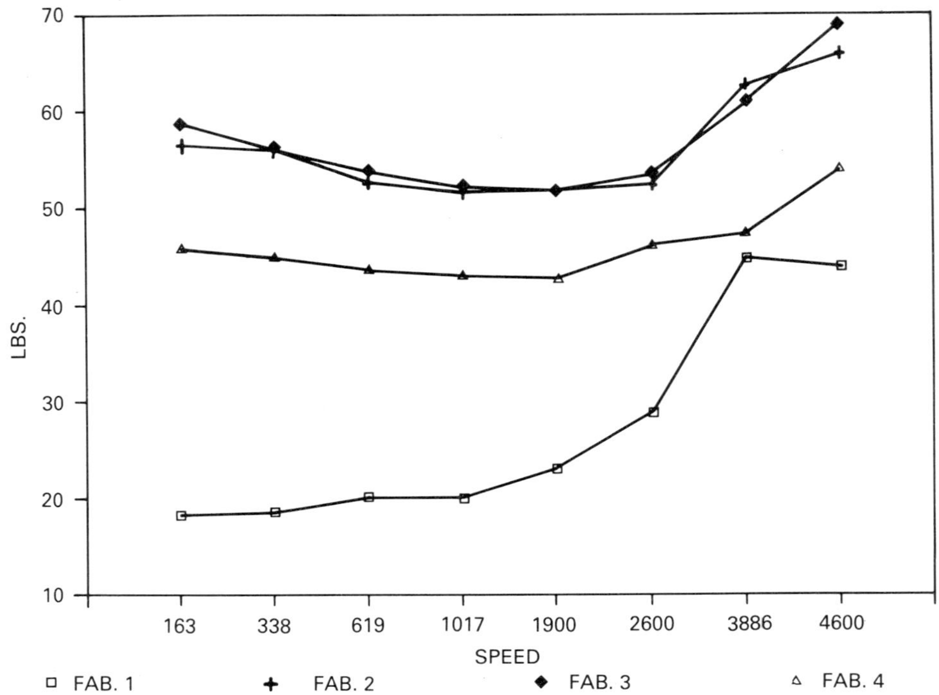

Fig. 5 — Combination 3 — range of forces exerted on the presser bar for eight levels of machine speed. Hinged (long base) presser foot used.

FABRICS AND THEIR IMPACT ON HIGH-SPEED FEEDING FORCES

Experimental design

A range of fabrics as described in Tables 1 and 2 were used for the correlation of fabric characteristics with feeding forces at high sewing speeds and 10 stitches per inch. All data was recorded at a sewing speed of 4588 rpm on the Pfaff 463 lockstitch machine. Data representing both the forces on the presser bar and its displacement was recorded simultaneously.

Displacement and feeding waveforms

Typical waveforms for the force and displacement are shown in Fig. 6 for both sewing combinations sewing fabric 1 from Table 2. Fig. 6(a) shows both the force and displacement waveforms for the double-hinged compensating and Fig. 6(b) shows waveforms for the hinged-spring presser foot combination. These waveforms show the different profiles for the same fabric and in particular the differences in the forces experienced by the two presser feet as the feed dog rises. It is also of interest to note that the displacement profiles show that the presser foot experiences first a large displacement followed by several smaller displacements during the stitch formation period. Each force waveform shows the large peak force at the end of the feeding cycle when sewing at 4588 rpm.

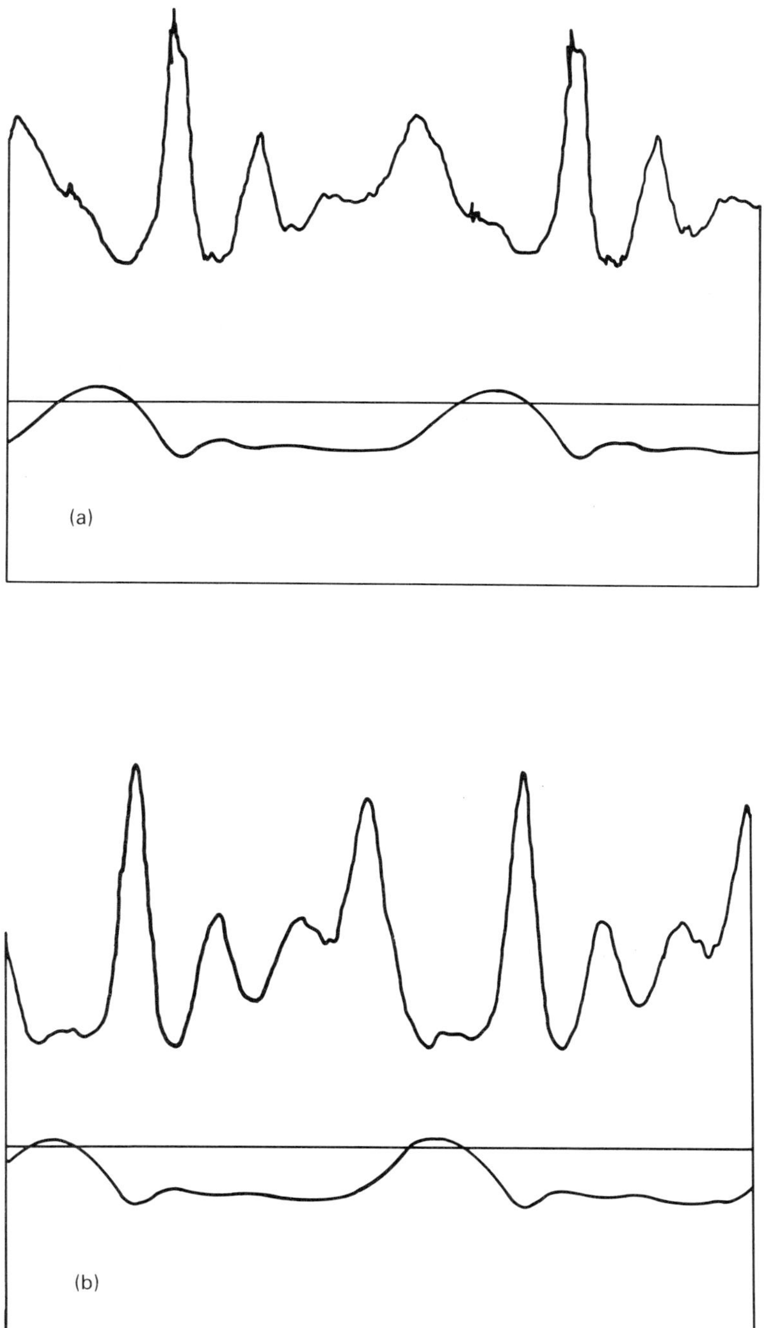

Fig. 6 — Displacement and force waveforms for: (a) fabric 1, foot 1 (double compensating hinged presser foot; (b) fabric 1, foot 2 (hinged foot).

The phenomena of presser foot bounce is well known and has been investigated by Johnson [15]. Figs. 6(a) and (b) show the extent of bounce for each of the presser feet under investigation. The stage of displacement beyond which bounce has occurred, shown as a horizontal line, is calculated as the sum of the feed dog height and the compressed thickness of the fabric being sewn. For the examples shown in Figs 6(a) and (b), bounce or feed dog and presser foot separation times were 0.0025 and 0.0018 seconds, respectively.

Results and correlation with fabric characteristics

The total force and displacement for one stitch were calculated for both the warp and weft directions of sewing. In addition, the sewn seam length, the relaxed sewn seam length, the amount of ply slippage and the thickness of the two plies were recorded. These 12 measures were correlated with the KES values according to the Pearson method of calculating correlation coefficient. The correlation matrices for warp and weft values are given in Tables 5 and 6 respectively showing only those correlations of significance.

Total feeding force

The total feeding forces for the double compensating foot showed a correlation with *LC* in both warp and weft directions. This is contrasted by the hinged-spring presser foot feeding forces correlating with *MIU, WC, B* and *T* in the warp and *EM%, MIU, WC, B* and *T* in the weft. This result suggests that, at elevated sewing speeds, the fabric property controlling the feeding force is different for different sewing combinations.

Total displacement of the presser bar

The total displacement of the presser bar correlates with *G, 2HG, 2HG5, B, 2HB, T* and *W* for both sewing combinations. The double hinged compensating foot also shows correlations with *WT* and *RC%* in the weft direction.

Thickness as sensed by the displacement transducer

The compressed thickness of the fabrics as measured on the sewing machine have significant correlations with the shear properties (*G, 2HG, 2HG5*), bending properties (*B, 2HB*), single ply thickness (*T*), weight and *MIU* for both presser feet and sewing directions.

Ply slippage

The double hinged compensating foot shows ply slippage correlating with *MMD, WC* and *T* in the weft and no significant correlations with the warp KES values. The hinged-spring foot trials show that ply slippage correlates with *EM%, WT* and *B* (weft only). It is again of interest to note that the different presser feet correlate with different fabric characteristics as far as the ply slippage is concerned.

Sample length changes

The percent change in sample lengths for the samples sewn with the double-hinged compensating foot correlate with the *RT%, G, 2HG, 2HG5, LC, RC%, 2HB, T* and

Table 5 — Significance levels for sewing parameters correlated with KES warp values

	EM%	*LT*	*WT*	*RT%*	*G*	*2HG*	*2HG5*
1Force							
1Displace					MS	MS	LS
1Thick					LS	MS	
1PlySlip							
1IntLeng							
1FinLeng							
2Force							
2Displace					MS	HS	MS
2Thick					LS	MS	LS
2PlySlip	LS		LS				
2IntLeng							
2FinLeng							

	MIU	*MMD*	*SMD*	*LC*	*WC*	*RC%*	*B*	*2HB*	*T*	*W*
1Force				LS						
1Displace							HS	HS	HS	HS
1Thick	LS						HS	MS	HS	HS
1PlySlip										
1IntLeng				LS						
1FinLeng				LS						
2Force	HS				LS		LS		MS	
2Displace							HS	HS	HS	HS
2Thick	LS						HS	MS	HS	HS
2PlySlip										
2IntLeng										
2FinLeng										

For 13 samples, a two-tailed test, the correlation coefficients are as follows: HS denotes high significance (0.0001)=0.8010; MS denotes moderate significance (0.01)=0.6835; LS denotes low significance (0.05)=0.5510.
1=double hinged compensating foot.
2=hinged foot.
Force=area under force curve.
Displacement=area under displacement curve.
Thick=thickness of fabric under the presser foot.
PlySlip=ply slippage of fabric strips.
IntLeng=initial length of fabric strips.
FinLeng=final length of fabric strips.

W in the weft but only *LC* in the warp. The changes in sample length for the hinged-spring foot correlate with *WT*, *B* and *W* in the weft and not warp fabric properties.

SUMMARY

This paper has presented some of the factors that need to be considered if full apparel automation based on objective measurement is to become a reality. Sewing machines generally must accelerate from idle to full speed for each operation. Data

Table 6 — Significance levels for sewing parameters correlated with KES weft values

	EM%	*LT*	*WT*	*RT%*	*G*	*2HG*	*2HG5*
1Force							
1Displace			LS		MS	MS	LS
1Thick			LS		LS	LS	
1PlySlip							
1IntLeng					LS	LS	
1FinLeng				LS	LS	MS	LS
2Force	MS						
2Displace					MS	MS	LS
2Thick			LS		LS	MS	
2PlySlip	LS		MS				
2IntLeng			LS				
2FinLeng			LS				

	MIU	*MMD*	*SMD*	*LC*	*WC*	*RC%*	*B*	*2HB*	*T*	*W*
1Force				LS						
1Displace	LS					LS	MS	MS	MS	HS
1Thick	LS						HS	LS	HS	HS
1PlySlip		MS			MS				LS	
1IntLeng				LS		HS				
1FinLeng				MS		MS		MS	LS	MS
2Force	MS				LS		LS		MS	
2Displace							MS	MS	MS	HS
2Thick	LS						HS	LS	HS	HS
2PlySlip							MS			
2IntLeng							MS			LS
2FinLeng							MS			LS

For 13 samples, a two-tailed test, the correlation coefficients are as follows: HS denotes high significance (0.0001)=0.8010; MS denotes moderate significance (0.01)=0.6835; LS denotes low significance (0.05)=0.5510.
1=double hinged compensating foot.
2=hinged foot.
Force=area under force curve.
Displacement=area under displacement curve.
Thick=thickness of fabric under the presser foot.
PlySlip=ply slippage of fabric strips.
IntLeng=initial length of fabric strips.
FinLeng=final length of fabric strips.

presented in this paper show that the sewing combination and the speed of operation influence forces on the presser bar and therefore the feeding of the fabric.

It has been shown that some fabric mechanical properties are correlated with the forces required to feed fabric through the sewing machine and the displacement of the presser foot. As in the findings with the study of machine speed, the type of sewing combination responds to different mechanical properties and in many cases the correlations are not sufficiently significant to establish adaptive control. This poses an enormous challenge for automation of the conventional sewing workstation.

REFERENCES

[1] CDI, Inc., Grand Rapids, MI, USA.
[2] Toray Industries, Inc., Tokyo, Japan.
[3] Sung Hoon Jeong, M. S. Thesis, North Carolina State University, Raleigh, NC, 1990.
[4] Dhingra, R. and Postle, R., 'Some Aspects of the Tailorability of Woven and Knitted Outerwear Fabrics', *Clothing Research Journal*, **8**, No. 2 (1980), 59–76.
[5] Kawabata, S.., 'The Development of the Objective Measurement of Fabric Handle'. In *Objective Specification of Fabric Quality, Mechanical Properties and Performance*, ed. S. Kawabata *et al.* Osaka, Japan: The Textile Machinery Society of Japan, 1982, pp. 31–53.
[6] Lindberg, J., *et al.*, 'Wool Fabrics as Garment Construction Materials'. *Journal of the Textile Institute,* **51**, (1960), T1475–T1493.
[7] Lindberg, J., *et al.*, 'Part III: Shearing and Buckling of Various Commercial Fabrics'. *Textile Research Journal*, **31**, No. 2, (1961), 99–122.
[8] Mahar, T., *et al.*, 'Fabric Mechanical Properties and Shape Formation in Suits'. In *Objective Evaluation of Apparel Fabrics*. Osaka, Japan: 'The Textile Machinery Society of Japan,. 1983, pp. 359–372.
[9] Mahar, T. and Postle, R., 'Fabric Equations for Australia, New Zealand, India and U.S.A.'. *Journal of the Textile Machinery Society of Japan*, **31**, No. 2, (1985), 35–38.
[10] Makabe, H. and Momota, H. 'Studies on the Deformation and its Direction of Skirts Caused by Wearing'. *Journal of the Japan Research Association for Textile End-Uses*, **27**, No. 9 (1986), 402–411.
[11] Matsudaira, M., *et al.*, 'Measurement of Mechanical Properties of Thin-Dress Fabrics for Hand Evaluation'. *Journal of the Textile Machinery Society of Japan,* **31**, No. 3 (1985), 53–60.
[12] Matsudaira, M. and Kawabata, S., 'An Analysis of Dynamic Drape of Woven Fabric'. *Journal of the Textile Machinery Society of Japan*. **39**, No. 12 (1986), T175–T183.
[13] Niwa, M. and Fusako, S., 'Relationships Between Drapability and Mechanical Properties of Fabrics'. *Journal of the Textile Machinery Society of Japan*, **39**, No. 11 (1986), 43–50.
[14] Saito, K., 'Fabrics Suitable for Sewing and Shaping for Wearing Apparels'. *Journal of the Japan Research Association for Textile End-Uses*, **25**, No. 2 (1984), 55–61.
[15] Johnson, E., 'Some Factors Affecting the Performance of High Speed Sewing Machines'. *Clothing Res. J.* **1** (1973), 3–35.
[16] Galuszynski, S., 'Comparison of Forces Opposing Needle Penetration Obtained with Medium Ball and Slim Set Needle Points'. SAWTRI Technical Report 571, 1985, pp. 1–20.
[17] Galuszynski, S., 'An Introductory Stuydy on the Effect of Certain Sewing Conditions on Fabric resistance to Needle Piercing', SAWTRI Technical Report 563, 1985, pp. 1–23.

[18] Hersh, S. and Grady, P., 'Needle Heating During High-Speed Sewing'. *Textile Res. J.* **39** (1969), 101–120.
[19] Horino, T. and Kawanishi, S., 'Simultaneous Measurement of Force and Temperature of a Sewing Needle Produced during the Needle Penetration Experiment at low Operation Speed'. *Sen-I Gakkaishi.* **32** (1980), T89–T95.
[20] Horino, T. and Iwamoto, S., 'An Instrument for Measuring Cloth Penetrating Behavior of Sewing Needles'. *J. Textile Machinery Society of Japan.* **26**, (1980), 55–57.
[21] Howard, M. and Virgilio, D., 'Sewing Needle Temperature — the Effects of Sewing Conditions', *J. Textile Res.* **43** (1973), 651–656.
[22] Howard, M. and Parsons, D., 'Sewing Needle Temperature'. *J. Textile Res.* **38** (1968), 606–614.
[23] Hunter, L. and Cawood, M., 'The Correlation between Two Different Sewability Tests and the Effect of Certain Wool Fibre and Fabric Properties of the Sewability of Woven Fabrics'. SAWTRI Technical Report 510, 1982, pp. 1–11.
[24] Hunter, L. and Cawood, M. P., SAWTRI Special Publication, Wol 50, 1979.
[25] Hurt, F. N. and Tyler, D. J., 'The Influence of Fabric Finishing Conditions on Sewing Needle Temperatures, *Clothing Res. J.* **1**, 47–52 (1973).
[26] Hurt, F. and Tyler, D., 'Seam Damage in the Sewing of Knitted Fabrics, I — Machine Variables', Hatra Res. Report 35, 1975.
[27] Hurt, F. and Tyler, D., 'Seam Damage in the Sewing of Knitted Fabrics, II — Material Variables', Hatra Res. Report 36, 1975.
[28] Hurt, F. and Tyler, D., 'Seam Damage in the Sewing of Knitted Fabrics, III — The Mechanism of Damage', Hatra Res. Report 39, 1976.
[29] Morooka, H. and Niwa, M., 'Physical Properties of Fabrics Relating to Making-up and Good Appearance'. *J. Textile Society of Japan.* **29**, 426–437 (1976).
[30] Nishikawa, J., Akashi, J. and Matsumoto, K., 'Study on the Differential Feeding and Seam Puckering of Knitted Fabrics'. *J. Japan Research Association for Textile End Uses.* **20**, 36–40 (1979).
[31] Onoue, M., Kinoshita, R., Kamata, Y. and Mikami, I., 'Received forces at Throat Plate, Feed Dog and Pressure Bar of the Sewing Machine for Industrial Use. *Sen-I Gakkaishi.* **36**, T354–T363 (1980).
[32] Osawa, M. and Saito, H., 'Measurement of Feed Dog Motion and Dynamic Torque in Sewing Machines by Means of an Optical Device. *J. Textile Machinery Society of Japan.* **29**, 66–71 (1983).
[33] Schwartz, P., 'Effect of Jamming on Seam Pucker in Plain Woven Fabrics'. *J. Textile Res.* **54**, 32–34 (1984).

QUESTIONS

STYLIOS: In the presser foot bounce graphs that you are showing, foot bounce is important; however, it depends upon the type of presser foot that you use. If you had a low inertia foot the graph will be different, so you would be able to definitely go beyond the 4000 rpm and have a totally different graph; thus one has to question which presser foot you are referring to in your diagram?

LITTLE: This is a double compensating presser foot, a regular off-the-shelf one.

STYLIOS: Yes, but it is not a low inertia foot; if it was low inertia you would get different results. You have a spring here not a cam.

LITTLE: No this is not a cam but it is a double spring system.

STYLIOS: Well, this is an improvement from the spring and not a low inertial foot as such.

LITTLE: This maybe so, but there are a lot of these around.

STYLIOS: I would like to comment that it is not true that sewing machines have not been instrumented before by means of having transducers attached to them. I can point out past work of a team at the Shirley Institute, who have attached strain gauges on the sewing machines to look at sewing performance, then the work that I did in the University of Leeds which is being continued here in Bradford, and of course the work at Denkendorf by Sandoz that has the same approach.

LITTLE: There are a number of systems around working with strain gauges and measuring forces on the throat plate, rather than in the needle bar or the presser bar.

STYLIOS: Our work in 1983 involved strain gauging the sewing needle itself for the first time to get the direct measurement of forces on the needle. The other question that I would like to ask is that having mentioned that you have attached a piezoelectric transducer on the needle bar, I suspect you had to modify the needle bar by cutting it.

LITTLE: Of course we have to cut the needle bar.

STYLIOS: Since you are looking at the dynamics were you able to compare your results here, because by cutting the needle bar, you had to retime the sewing machine, consequently the dynamic measurements, I assume, would not be the same.

LITTLE: No, that is why I did not touch on that today, but it depends what you pick up; you pick up all the inertia of the needle and the needle bar piece that is below, then you must subtract all the inertia forces so that you can get your pure sewing penetration and witrhdrawal forces and that is relatively easy to do in your calculation of wave form. Yes you are right but you can subtract it.

STYLIOS:The other thing in the graphs that you've shown, especially when you measure the forces on the presser foot, it is clear to me that the trace is different to the trace that we are getting. May I therefore ask what was the accuracy of the measurement of data? How many points are you reading per cycle because this is significant?

LITTLE: We standardize for about 300 points.

STYLIOS: Then what you are getting is a presser foot bounce as well as inertia and vibration effects, and if you work back to the first engineering principles different fabrics have different damping characteristics, so you can relate that to the fabric properties but the idea here is that really from the engineering point of view you have to actually damp the system, because it is the engineering of the system that is wrong, i.e. if you can damp the presser foot then you do not have this problem, it is the damping action of the fabric and if I can go a little further we have actually developed a damping mechanism based on a piston working with air or oil, which acts on the foot and eliminates bounce, vibration or inertia effects. So these are known curves between the two foot bounce graphs that you have, which are nothing else than vibration and inertia effects, due to your method of measurement.

LITTLE: That is true and a number of sewing machine companies have developed reasonably sophisticated systems on the presser bar when it comes to high speed. We are only working at 4500 rpm, others are working at 9000 rpm and this is a differennt system. We have the capability of getting up to 9000 rpm but this for us is not the highest priority, because it is very difficult to get funding for this work.

POSTLE: You've mentioned the way you are analysing your wave forms I am just wondering whether you've got an idea of the basic frequencies involved in your variations.

LITTLE: When we go through say a Fourier transformation, most of those in the needle bar go out to about 100 kHz in frequency.

POSTLE: In our work, we have also noticed very high frequencies of that order so I try to comment on the importance of the dynamic measurement as opposed to static measurement in this context when you are talking about 100 kHz. I would just like to refer to my very last overhead because these measurements are quite easy to make and pretty consistent: the sonic technique that I have shown on the dynamic modulus measured on fabrics. One could take this sort of measurement which is very rapid and easy to do instead of the static measurement.

LITTLE: I totally agree with you, I think it is the type of work that we should encourage jointly as we go forward, because I think it will be very exciting to find out if there is a relationship to some of those sewing characteristics.

POSTLE: A comment on the question you have asked the delegates on the difference between the warp and weft thread. It seems to me that this could be a simple experiment in a sense that you only measure and use correlation, bear in mind that these objective measurements, shear, bending, etc., are not completely independent from each other; there is a correlation between them and so when one does regression or whatever, you can get say the shear properties coming out as the most important property variation in the warp, and in the weft you may come up with the bending property as a number one, but remember that there is correlation between them; it's a little bit which one comes out best. There may not be any basic significance.

LITTLE: I agree and one of the things we present here is the full correlation matrix because many things correlate highly with each other. We are aware of that and we would prefer to have a more scientific relationship, and I think this is the direction we want to pursue rather than a correlation.

AMIRBAYAT: There was not enough data on the force transducer on the needle bar.

LITTLE: I did not show any data, this is ready to be written up, I did not know if we had time to go through that and we've done different trials on different needles and threads and balancing stitches. It is more appropriate for thread, sewing needle, lubricant manufacturers and those involved with the setting and the dynamics of sewing machinery, rather than objective measurement.

AMIRBAYAT: How do the needle and presser bar data compare?

LITTLE: They are completely different; the forces on the needle bar are much lower, because you just penetrate fabrics, you get complications because of the amount of damage you do to the fabrics, this is also complicated by the thread diameters and how the thread bends as you go through materials and what you are measuring.

DeARAUJO: Just a comment about the damping variations mentioned earlier. As I understand it, if you have sewn through paper and rubber and distinguish what the variation is, then by this standard method you can establish where your variations are coming from.

LITTLE: Yes, previous work that is published looks at all different types of material, and in the case of rubber there are none of these forces at all; there is a simple straight line; the paper is almost like having nothing at all, i.e. getting presser foot forces, and fabrics are all in between.

DeARAUJO: This is a useful technique used to understand the data.

LITTLE: I agree, but I think how you fingertip those wave forms can provide a better understanding on these characteristics.

Introduction to Dr M. Sarhadi and J. L. Catchpole

The next presentation is work based at the University of Durham from Dr Mansoor Sarhadi who is in fact from October moving to Brunel University as Professor of Manufacturing and Engineering Systems. He will be starting the presentation which will be followed by Mr J. L. Catchpole his Ph.D. student who was awarded his first degree as well as his M.Sc. from Durham; and he is to complete his Ph.D. in October. The title of the paper is 'Stitch quality monitoring in sewing operations'.

Stitch quality monitoring in sewing operations

J. L. Catchpole and **M. Sarhadi**
University of Durham, UK

ABSTRACT

The *tension* on the two *needle threads* and the *looper thread* of an industrial *chainstitch sewing machine* have been monitored using *strain gauges* and the data recorded using an analogue-to-digital converter and 68020 based microprocessor. The data collected has been analysed to identify the variations relating to *missed stitches*. Three different techniques have been applied to the *automatic detection* of these variations. This analysis indicates thus far that detection of mis-stitches is possible and a viable *real-time algorithm* is being developed.

1. INTRODUCTION

The development of the modern sewing machine has been conducted in a largely ad hoc manner over the last century or so. The basic lockstitch and chainstitch mechanisms have changed little over the last 60 years or so [1,2], the only real developments in this period being the improvement of the construction and lubrication of the machines resulting in ever higher operating speeds. Modern sewing machines can now run at speeds of up to 10 000 stitches per minute and operating speeds of 6000 to 8000 stitches per minute are not unusual. Such machines can accelerate to full speed from rest in less than half-a-dozen stitches and the most modern can do so in less then one stitch.

With these machines being run at such speeds and in a stop–start manner as garments are fed in, problems are common. Damage to the material being sewn or to the threads used is impossible to avoid and is tolerated so long as it does not cause breakdown of the seam. Puckering of the seam is a common problem [3,4]. Thread breakages are frequent and irregular stitch quality along seams often has to be accepted. These problems have already been studied intensively and guidelines established to deal with most of them drawn up [5–17], although predicting their occurrence is difficult or impossible [3,5,7,18].

One important problem has been left largely unstudied and has proved difficult to deal with. This is the occurrence of mis-stitching or skip-stitching, in which every so often a machine which is otherwise behaving acceptably will miss out one or more stitches in the middle of a seam, then pick up the pattern and carry on as if nothing had happened.

This problem is endemic to the sewing industry and guidelines on what may be causing it and how to deal with it appear in the trouble-shooting and maintenance sections of both industrial and domestic machine handbooks [19,20].

The usual approach to dealing with the problem in industry is generally compensatory in nature. As long as no single machine is mis-stitching too often, the only action taken is to have experienced staff check the seams produced for these faults, either when checking for other problems (oil-stains, torn panels, etc.), or on a statistically based sampling of the in-process work. Mis-stitches, like other stitching faults, are correctable [6,7], the faulty seam being unpicked and re-sewn. Apart from the extra thread and possibly some binding tape, the only cost arising from the occurrence of these faults is the cost of this quality control, which takes up a lot of time and effort [6,7].

If a machine starts producing too many faults, then a mechanic is called to service it. The usual corrective action is to check the timing and the tension settings on the threads and if a fault is not found, replace the thread bobbins (in case they are poorly wound or the thread is of poor quality). If this does not correct the problem, the machine is taken for overhauling [22].

A common culprit is the thread. Since the characteristics of the thread and the thread/fabric combination are important [3,22,23], particular threads can cause problems in some applications, often for no obvious reason. Some work on the reasons for this is known to have been carried out and published [23], but most knowledge appears to be empirical [3,4].

The sources available from manufacturers, that is machine operating handbooks or general applications guides [19–21], quote a number of known problems, including misalignment of the needles or the loopers, damage to the needles or loopers arms, wrong size of needle or thread and poor timing of the machine operation. These are however, only given as reasons for excessive mis-stitching, not mis-stitching in general.

While a large amount of research has been and is being carried out on quality control at all stages of garment production, from the properties of the threads and their dependence on the manufacturing techniques used [40–44], to the finished garments [6,7], most research in the specific area of seam formation seems from the published work to have been concentrated in studies of physical seam strengths and properties, comparing mostly the single-thread chainstitch, the two-thread chainstitch and lockstitch and the three-thread overlock stitch (e.g. [24–26]). These studies covered the dependence of seam strengths resulting from the formation of the various seams [27,28]; the heating effects on the needles and the possible consequent damage to the threads and fabric [10]. Such studies still appear to be under way in some parts of the world [9,10].

The studies of dynamic tension in the sewing threads which have been conducted have generally been intended as a research tool only. They have mostly been aimed at finding quantitative estimates for the stresses suffered by the threads and later the variations introduced by charges in operating conditions [10,29–34]. In this country, identifying indvidual faults in a seam does not seem to have been an aim of any but the latest studies, which attempted to diagnose mis-stitches from the tension data [36,37]. These were abandoned, apparently largely to the enormous problems inherent in collecting the necessary data with the technology available, which was based upon photographic recording techniques [22]. One Japanese study is known which attempted to establish a basis for mis-stitch detection using tension data [35]. This work appears to have been discontinued.

The problem of mis-stitching is one which has to be solved to enable the future automation of the garment industry, when human monitoring of sewing workstations will be impossible both from a practical point of view and for cost-effectiveness.

3. DATA COLLECTION

The work covered in this paper has been aimed specifically at developing a means for identifying mis-stitches as they occur. Depending on the cost, the resulting system could be used as a quality control device for existing machines, as a diagnostic system for machine and thread trouble-shooting or as a monitoring system for such future automated systems.

The first part of this work was to correlate tension data to physical seams and thereby identify the changes in the data which are caused by sporadic faults. The various kinds of mis-stitch are the most important of this kind of fault. Other problems are largely those of poor machine set-up or long-term faults. These should show up as general distortions in the data, although it has been found that on some machines good seams can be produced from apparently poor profiles, a finding backed up by earlier work on lockstitch machines [18].

The sewing machines used in this study were a pair of Rimoldi B63 chainstitch machines with two needles, producing a double line of stitches with a common looper thread (seam no. 406, see Fig. 1). The double line of stitches is intended to be

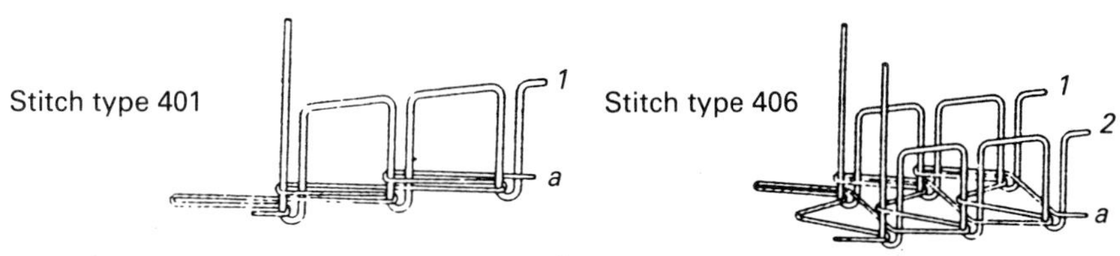

Fig. 1 — Seam numbers 401 and 406.

stronger than a single 401 seam, but to otherwise retain the same characteristics. It is therefore ideal for the high stress, high strain seams required in undergarments, particularly since its more obvious appearance is of less importance on a concealed garment. While this is an older machine, it is reliable and does not possess any of the complex control electronics of the newer machines. Tensioning is accomplished on all three threads by using a single pair of sprung discs with, in the case of the looper thread only, a second small disc pair at the end of the thread path. The two needles are at different heights, the one furthest from the looper being the lower by about $\frac{1}{4}$ inch. This is apparently to ensure that the two threads do not go slack at the same time and foul each other.

The machine was fed a 1.25 inch knitted cotton cloth tape of the sort used for binding the edges of the material on undergarments. The threads used were a mixed group, APEX 60 cotton thread, KOBAN 120 corespun polyester/cotton (61%/39%), Donisthorpe's Don Blend Soft, Courtauld's Primax 120 and BST's Zeta with SAF finish.

The tension in the three threads has thus far been monitored using a set of cantilever arm strain gauges, each with four resistance gauges mounted in two pairs on opposite faces of the arm. These elements are connected electrically in a bridge circuit. The gauges have been used for data collection at stitching speeds of up to 2900 rpm, which is the limit for the drive currently in use with the research machines.

These gauges pick up any stray vibrations from the general environment, so the gauge arms are mounted on a frame sited on the sewing machine table. Since the sewing machine is mounted on the table using four large rubber feet thus providing very effective mechanical decoupling of the vibrations generated when it is operating, this isolates the gauges from these vibrations. Previous tests had shown that the machine generated unacceptable levels of vibration at stitching speeds of around 1500–2000 rpm and higher levels at higher speeds. A pair of guide arms fixed to the mounting frame and sited in the thread paths provide the control of the threads over the gauge arms (see Fig. 2). While this approach works as a research tool, it is too cumbersome for an industrial system, so an alternative is needed.

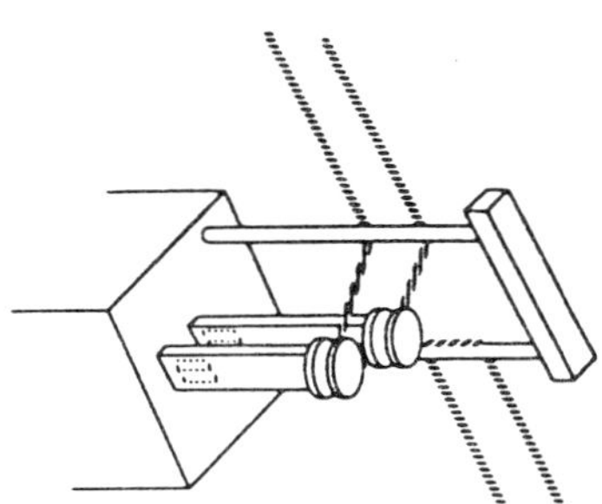

Fig. 2 — Mounting of strain gauges.

The use of resistive strain gauges also has other problems. For instance, to avoid spurious resonances the gauge arms had to be designed so that their lowest natural frequencies were above the highest frequencies in the thread tension signal. Also, resistive strain gauges have a frequency limit of around 9 kHz and attenuate all signals above this. This is currently less of a problem since the useful information is concentrated at the very bottom of the spectrum and losing the higher frequencies is no problem. However, at very high machine speeds (8000 rpm) this would be unacceptable. Furthermore, the signal from these gauges is very weak and requires a very high (around 300 times) initial amplification using a precision instrumentation op-amp to give a useful signal.

For these reasons a new gauge set based on piezo-electric ceramic gauges is being built. These have a much higher natural frequency, give a stronger output signal and do not require a stable voltage source to drive them, as do resistive gauges. The signal from the gauges is fed to a four-stage 8th order active filter with a small Q value, to give a smooth roll-off. This has a corner frequency of 12 kHz and an attenuation of 45 dB at 22-kHz.

The data is collected using a 68020 based system with an AD7572 ADC chip. The AD7572 has a minimum settling and conversion time of 8 μs, giving a nominal maximum sampling frequency of 125 kHz. This must be divided by the number of channels used, which at present is three (the two needle threads and the looper). Thus the current maximum sampling frequency is around 42 kHz.

The 68020 host board has 1 Mbyte of system memory, of which 809 kbyte is allocated for temporary data storage. This allows storage of a total of 414 000 samples, or 138 000 per channel. The other data stored is derived from pulses generated by an opto-switch connected to a slotted disc on the drive shaft of the machine. These pulses are generated once every stitch cycle, when the needles are just at their highest, and are fed to the 68020 via a serial port. The 68020 records the number of samples between pairs of these pulses thus giving an index of the lengths of the individual stitches. This gives an easy way of breaking up the continuous data stream into meaningful stitches. This gives an easy way of breaking up the continuous data systems into meaningful segments.

The noise level on the ADC inputs from the system as a whole (including ADC noise) is around 34 dB. This would be a high noise level for audio processing but here the noise is barely discernible.

3. DATA ANALYSIS

The formation of the 406 chainstitch is a complex series of actions and before the importance of the tension profile data can be properly grasped, it must be understood. The seam is based on two parallel 401 chainstitch seams (see Fig. 1) using the same looper thread. The easiest way of considering the formation of this seam is to look at the formation of a 401 stitch (see Fig. 3)).

The formation is difficult to describe easily, since it is a continuous process with no distinct start or stop points. To pick an arbitrary starting point, consider the needle retracted fully (Fig. 3(c) and (d)). At this point in the operation, the looper is towards the front of the machine and is fully across under the needle. It is holding a loop of the needle thread from the previous stitch which has been stretched from the

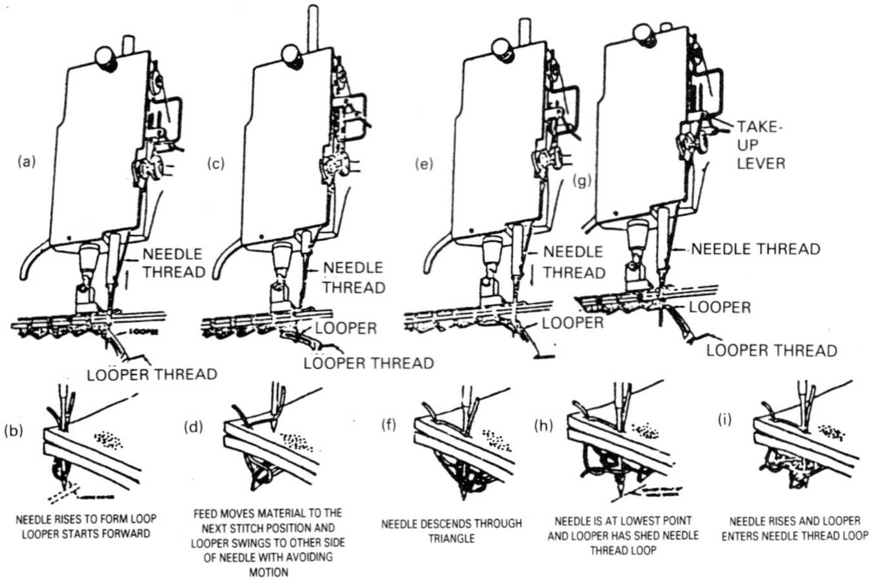

Fig. 3 — Formation of the 401 seam.

previous stitch on the material to the looper's current position. The looper is also trailing its own thread from the previous stitch. The needle descends and passes inside the looper thread and to the left of the loop of needle thread, as viewed from the front of the machine (see Fig. 3(e) and (f)). The looper then retreats and drops the loop of needle thread it was holding, then rotates to the back of the needle (see Fig. 3(g) and (h)). The needle then rises, tightening the released loop of needle thread. As the needle rises a loop trails the needle eye. The looper moves back in and passes through this loop (see Fig. 3(i)). The needle finishes rising and the feed dog then advances the material. Finally, the looper rotates towards the front of the machine, coming back to the start point.

In the 406 seam, this situation is complicated by the presence of the second needle thread. When the looper comes forward, it passes through trailing loops of thread from both needles. Hence when the needles retract, two loops of needle thread are held by the looper arm. When the needles descend, the outer needle penetrates the material first. It passes through the triangle formed by loop of its own thread and the looper thread. The looper retreats slightly, releasing that loop but still holding the loop of thread from the other needle. The second needle, which is set slightly higher than the first and thus penetrates the material after this has taken place then arrives. It passes through the triangle formed by the second loop of thread and the looper thread and the looper then releases the second needle thread loop. Thus from the point of view of the needle threads, each forms a 401 seam with the looper thread. The looper thread, however, loops around both of the needle threads in each stitch and therefore forms a more complex pattern (see Fig. 1).

Examining the 401 seam, it is obvious that there are two points at which a mis-stitch could occur. These are when the needle passes through the triangle formed by the loop of needle thread and the trailing looper thread (Fig. 3(f)), and when the looper passes through the loop of needle thread trailing from the needle eye as the needle rises (Fig. 3(i)). For the remainder of this paper, these shall be referred to as Class 1 and Class 2 mis-stitches respectively.

On the 401 seam, these two mis-stitches lead to similar results. If a Class 1 fault occurs, then the loop of looper thread from the previous stitch is released. If a Class 2 fault occurs, then the needle thread is not pulled to the following stitch and does not engage thread there. In effect, the loop of looper thread to the following stitch is released.

On the seam, however, the results are different because of the presence of the second needle thread. If a Class 2 fault occurs on either needle thread, the results are very simple. The trailing loop of needle thread is dragged back through the material as the needle rises and the previous stitch tightens. If the seam is examined at this point it is clear that a stitch is absent (see Fig. 5).

The result of a Class 1 fault depends on which half of the seam the fault occurs. If a Class 1 fault occurs on the other thread, a loose loop of looper thread results. This is still held by the previous stitch which has already tightened (see Fig. 4(a)). Over a

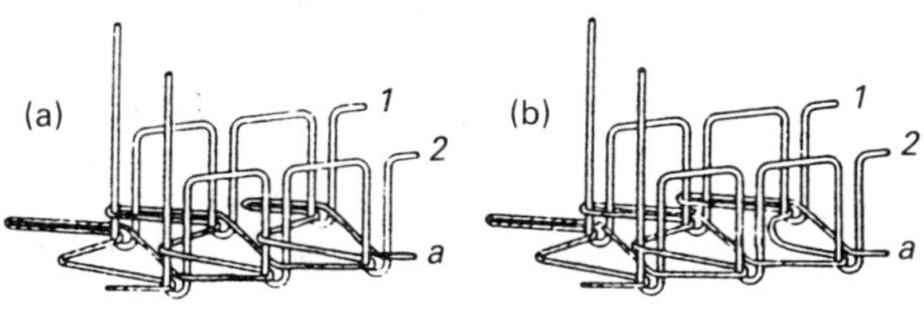

Fig. 4 — Mis-stitch Class 1 examples.

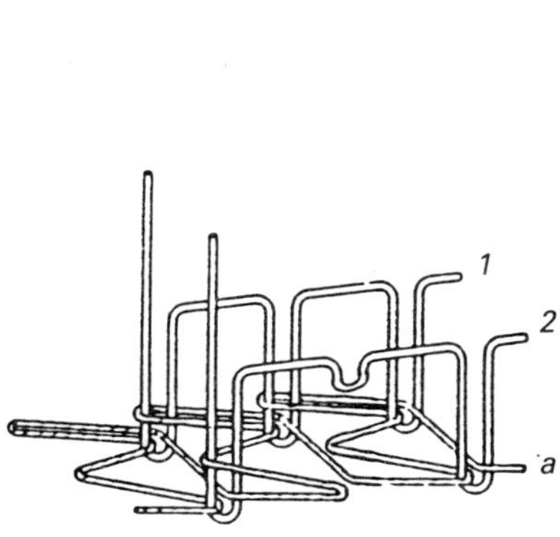

Fig. 5 — Mis-stitch Class 2 example.

period of time, this loop may loose, leading the stitch to unravel as if a Class 2 fault had occurred. If a Class 1 fault occurs on the inner thread the looper thread is still held by the other needle thread. This leads to an improperly formed stitch (see Fig. 4(b)).

From the above, it could be deduced that a Class 2 fault would affect the tension on the needle thread involved to some degree, but not the looper thread since the looper arm is involved in the formation rather than the thread. Using similar reasoning, the Class 1 fault would tend to affect the looper thread tension rather than the needle tension for the opposite reason.

In this case, the Class 2 fault is the worse of the two, since for the duration of the fault, the seam is reduced to half strength (only one needle thread is holding the seam). Since a seam will only be as strong as its weakest point, this is a serious problem. The Class 1 fault is much less important. While the stitch is complete, most of the structural elements of the pattern are still present. On stretching the material, the Class 1 fault behaves similarly to the rest of the seam and does not gape in the same way as the Class 2 fault. This fault has only recently been encountered, previous information suggesting that is was unlikely to occur.

Close examination of the tension profiles for the sewing cycle on the two needles, reveals the following points. The profiles for the two are radically different, reflecting the effect the combination of the two in one seam has on the forces involved in the formation of the seam. Comparing these to the profile for the simple 401 seam [32], it is found that this profile is almost identical to that for the outer thread. From this it can be deduced that the presence of the second thread is generally unimportant to the formation of this half of the seam.

The second profile, for the inner thread, is very different, the peaks present in the middle of the profile, where the stitch formation takes place, are much smaller. This indicates that the higher peaks in stitch formation must result from interaction with the looper, since the only difference between the two threads is that the outer thread is already holding the looper thread when the inner thread engages it and is still holding it when the inner thread releases it. The looper profile is considerably simpler than the others, having only one peak of any size.

Once the above has been considered, relating the features of the profiles to the action of the stitch formation is quite straightforward. The single peak on the looper thread profile (see Fig. 6) matches the point at which the thread needles have risen and is due to the retraction of the looper arm when the needle threads are taut. The

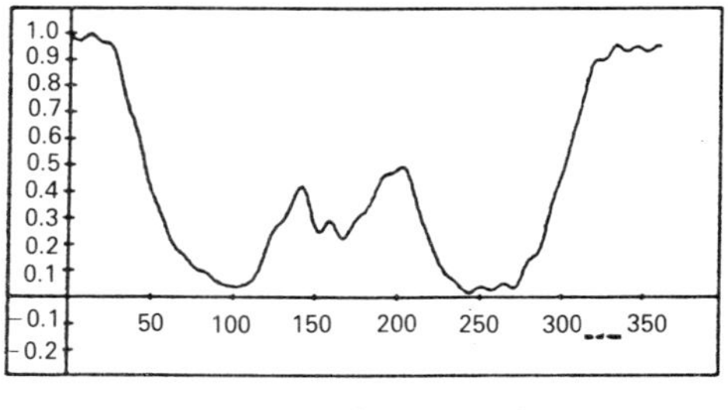

Fig. 6 — Outer thread profile.

looper thread is left to hold the needle while the looper arm moves back quickly, pulling on the looper thread. Thus its tension jumps sharply. It falls as the needles come back down, letting the tension in the needle threads drop. For the remainder of the time, while the looper arm holds the needle threads, the looper tension is close to zero. A very small secondary peak exists which has a characteristic shape, but varies so much in normal stitching that it indicates nothing useful. The height of the main peak varies with the setting on the two thread tensioning discs, much as the main peaks in the needle threads do. Unfortunately the presence and action of the looper arm mean that there is very little useful information in this profile.

Figs 7 and 8 represent one cycle of stitch formation on the outer and inner threads respectively. Each starts and finishes with the needles at their highest point and covers one revolution of the sewing machines drive shaft. The two needles drop

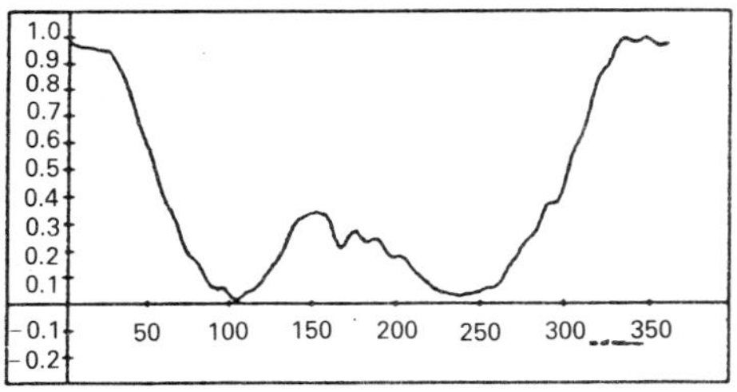

Fig. 7 — Inner thread profile.

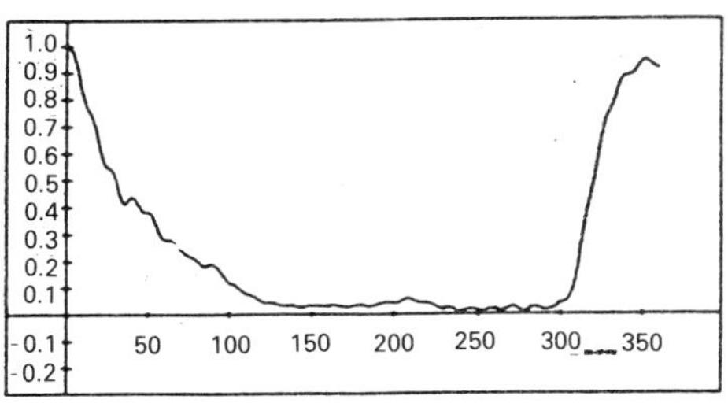

Fig. 8 — Looper thread profile.

together and as they do so both threads go slack. The outer needle reaches the material at about 80 degrees, the inner at about 90 degrees, the difference reflecting their staggered heights. The effects of this are seen after a short delay, the tension beginning to rise as the slack is taken up below the material. The needles reach their nadir at around 180 degrees, but before this happens, both needles have passed through their respective triangles of thread and the looper has disengaged from both of them, causing a sudden drop in tension.

On the outer thread, the tension begins to rise as first the retreating looper tightens the loops, then the inner needle hooks the looper thread and tightens it. The tension on the inner thread does not display such a pronounced second peak since the looper tension has already reached a maximum by the point when it begins to tighten. The two tensions then fall as the needles start to rise, slackenning the threads. The looper arm hooks first the inner then the outer threads and from this point on the tensions rise as the thread is held taut between the needles and the looper arm. It is at this point that the extra thread for the next stitch is pulled through the tensioning discs by the tensioning arm.

The heights of the main peaks and of the second of the two smaller peaks are dependent on the needle thread tension, while that of the first of the two smaller peaks is dependent on the looper thread tension. No correlation between the tensions of either of the needle threads and the profile of the other thread is observed, as would be expected. The balance of the seam with respect to the looper and either thread is therefore indicated by the ratios of the heights of these three peaks. This is not as straightforward as the above might indicate, since there is a large variation in the heights of the aforesaid peaks, even for the good seam they indicate. Averaging these heights might help, except that the Koban 120 is prone to spikes on the second of the smaller peaks, apparently the result of snags caused by the more frequent knots on this thread. The Apex 60 does not seem to suffer from these sudden spikes, although the smaller peaks are less regular. Reversing the two threads confirmed that it was the threads and not the tensioning gear causing these problems. The exact cause of these spikes is difficult to determine, however, when the Koban was run through the machine on all three paths, they did not appear to occur, thus it may be a thread to thread interaction with a looper thread (the BST Zeta was used for the looper in these tests).

Comparing the different threads at the same tension settings, both the profiles they lead to and the seams they produce were found to vary widely. To study this, the machine was first set up with the Koban 120 on all three paths and the tensions balanced to give a good seam. The profiles for this were recorded, then the Koban was exchanged for the Coutaulds Primax without changing the tensions and the profiles again recorded. This was repeated in turn for the Don Blend Soft, Apex and BST threads. Comparing the results, none of the threads produced a good seam except the Koban, for which the tensions had been set. All of the others produced a number of Class 1 mis-stitches and the Don Blend gave two Class 2 mis-stitches, which unfortunately occurred outside the recorded section. The profiles varied most noticably on the outer thread, the primary peak dropping by around 15% on the Coutaulds and by about 5% on the Don Blend Soft. This peak rose by similar margins on the BST and the Apex respectively, becoming very irregular in overall height on the BST. The first of the secondary peaks fell on all but the Don Blend, where it remained constant. The greatest change was on the second of these two peaks. Although in the Courtaulds profile, this peak did not change appreciably, in the case of the Apex, it remained about the same but became very irregular, while for the BST it rose around 50% and became very irregular. The Don Blend profile changed radically since this peak doubled in height and thus topped the primary peak. Previous attempts to generate mis-stitches by unbalancing a seam indicasted that such a profile was symptomatic of very badly balanced seam, prone to mis-stitch, which agrees with the quality of the seam produced. The variations in the BST profiles suggested that the tensions established were slightly too high on the outer thread and too low on the looper. The Courtaulds appeared too low on both threads, while the Apex thread seemed too low on the looper. The Don Blend appeared to be ambiguous since the two peaks dependent on the needle thread tension vary in opposite directions.

Taking the remedial action indicated by the profiles was found to give a better seam. In the case of the Don Blend the required remedial was found to be raising the tension on the outer thread.

The tension profiles for the two classes of mis-stitches are very different. In the case of the Class 1 mis-stitch, as expected little is to be seen on the needle thread profiles, certainly nothing which can be reliably used as an indication of the fault. Unfortunately, little change is found in the looper tension either, since one of the needle threads is holding the looper thread at the crucial time.

The Class 2 is a different matter. Whilst, as expected, there is little or no change in the looper tension, the 'average' Class 2 fault gives a needle thread profile like that in Fig. 9. This shows the profile for three machine cycles, the fault being revealed at

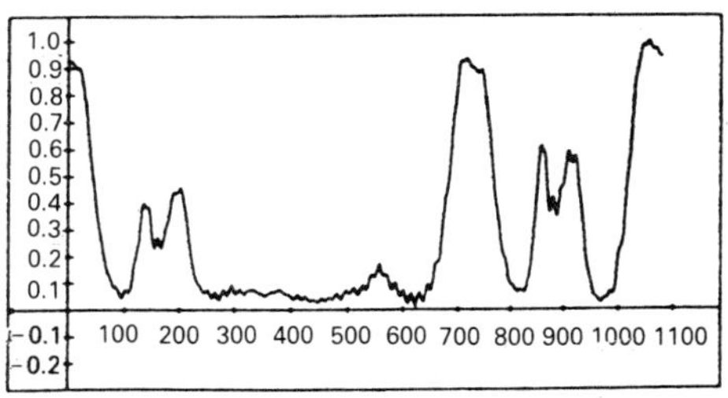

Fig. 9 — Obvious Class 2 profile.

about the 250 degrees point, where the looper arm should have caught the needle thread loop on the first stitch cycle. As a result, the thread remains slack as the needle rises, the thread loop having been pulled back out of the material. The small peak between 500 degrees and 600 degrees suggest that the needle engaged the looper thread properly and the following large peak as the needles withdrew for the second time indicates that the looper picked up the needle thread correctly this time. From here on the seam is normal.

This kind of deviation in the profile is easy to detect. However, the change is not always so obvious. Profiles have been collected for the same faults with much smaller changes (see Fig. 10) which are believed to arise from situations where the looper engages the thread, but it slips at some point as the needle is rising and the thread goes taut. This gives the kind of profile where the main peak exists, but is distorted as in Fig. 10 between 250 and 400 degrees. The following secondary peak is smaller than normal, due to the slackening of the thread caused by the fault. These situations have been found to be much less common than the clear fault.

4. AUTOMATION ALGORITHMS

A full mis-stitch is thus normally obvious in the tension profile and the need is for a simple robust algorithm to do so. The complications are the presence of vibrational noise from the sewing machine when it is running at speed, distortions caused by minor problems such as momentary snags in the thread, and considerable variations in the normal profiles from stitch to stitch. The analysis must be carried out on each stitch made by a sewing machine running at 6000 rpm, or more, which allows less than 10 ms to fully process a stitch and generater the diagnostic.

The three main approaches which have been tried are based on Fourier analysis, autocorrelation and adaptive liner prediction [46,52,53].

Fourier analysis is a technique for generating a representation of the frequency content of a signal. The result of the transformation is a set of coefficients each of which represents the strength of the signal at a particular frequency. The frequency range of the allowed components is limited to half the sampling frequency, a result of representing the waveform by a set of samples.

This transformation is useful since variations in a waveform are often more easily observed in the changes in its frequency content than in the waveform itself. For more detail see [54].

Good stitches were found to give a spectrum with a pronounced peak at very low frequencies, which appers to carry most of the information and a single secondary peak which is about 15 to 20 dB lower than the main peak (see Fig. 11). The large variations in the tension profile due to a full mis-stitch were found to lead to a drop in the main peak of around 5 dB and are therefore detectable. The spectrum is otherwise basically unchanged. More subtle variations arising from Class 2 mis-stitches tend to distort this primary peak and can therefore in principle be picked out.

The most important problem in this application is that resulting from the representation of the spectral content of the signal only by a set of discrete frequencies. The components at other frequencies must be represented by the allowed coefficients, which tends to spread the spectral peaks. This makes resolving the very bottom of the spectrum difficult. The number of spectral lines calculated is fixed by the duration of a single stitch (which determines the number of samples involved in the transformation). To improve the resolution, two or more stitches can be taken at once, but if a good stitch is included with a mis-stitch, particularly a subtle one, the distortions are reduced. Furthermore, the time required to calculate one transform of twice the length is greater than that to calculate the two individual transforms. It is possible to improve the resolution by adding null data to the input ('zero padding') but this leads to much greater calculation times since to avoid distorting the pattern of the spectral coefficients, the number of data values must be increased to at least double the initial number. Thus there is a trade-off between the resolution of the transform and the amount of calculation required.

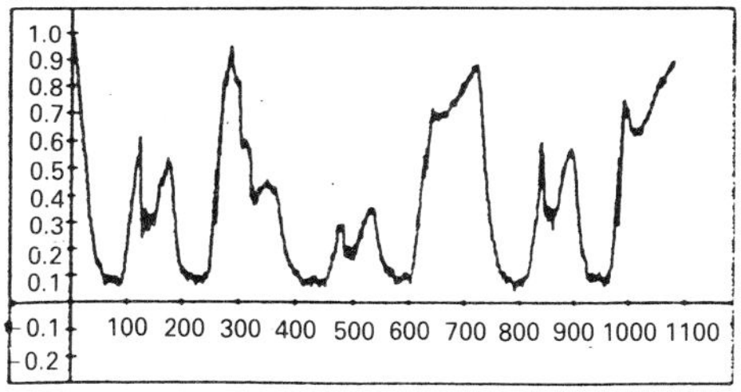

Fig. 10 — Subtle Class 2 profile.

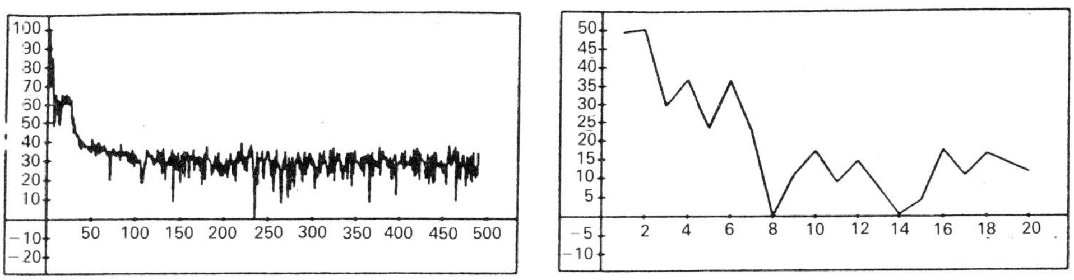

Fig. 11 — Good stitch spectrum.

As in the original tension profile, lesser faults do not appear to lead to any identifiable variations in the spectrum. Any variations which may exist are overwhelmed by the variations which occur from stitch to stitch in normal operation.

The well-known techniques of linear prediction are widely used data compression, modelling and noise reduction. The linear predictor estimates the required output by using a weighted series of the past input data values, a process very similar to an FIR digital filter. Taking the estimate $y(n)$ produced by the predictor as

$$y(n) = \sum_{k=0}^{M} h(k).x(n-k)$$

if the correct value for the next member of the sequence is $y(n)$ then the error, $\hat{e}(n)$, in the predicted value is

$$\hat{e}(n) = \hat{y}(n) - y(n)$$

The techniques of adaptive linear prediction attempt [40] to minimize this error according to some criterion by adjusting the predictor weights, $h(k)$, to give a better estimate of the next value. One common criterion taken is the minimization of the mean squared error (mse). This is a quadratic function of the predictor weights and therefore has a unique minimum. Thus the minimum can be found by decending the gradient of the function. A number of algorithms exist to increment the coefficient values towards this minimum, the best known of which is the LMS algorithm due to Widrow and Hoff.

In the LMS algorithm, the gradient of the function is estimated as

$$g_k(n) = -2.\hat{e}(n).x_k(n)$$

An increment proportional to the negative of this estimate is added to the coefficients, i.e.

$$h_k(n+1) = h_k(n) + 2.\mu.x_k(n).\hat{e}(n)$$

The constant, μ, is known as the convergence coefficient and is crucial to the behaviour of the predictor in any application.

The work has so far used the LMS algorithm. While this algorithm is not always the most effective, compared to others [54,55], it is the simplest form a computational point of view and is generally better understood [47,49–52,55].

The application requires the detection of variations from the norm in the data. Therefore, the predictor estimates themselves were not considered as the output, but rather the errors in them were analysed and also the scale of the weight adaptation (see Figs 12 to 14).

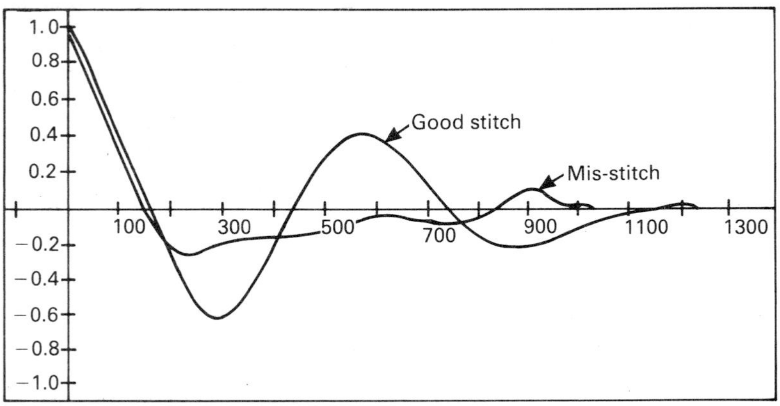

Fig. 12 — Autocorrelation functions of good and bad stitches.

The data was fed to the predictor as both reference and input with an offset of one 'stitch' period. This gave the difference between the signal value for the stitch and the current estimate as the error, in effect using the predictor to cancel the good stitch data leaving the variations from it as the output signal. The problem then is keep the

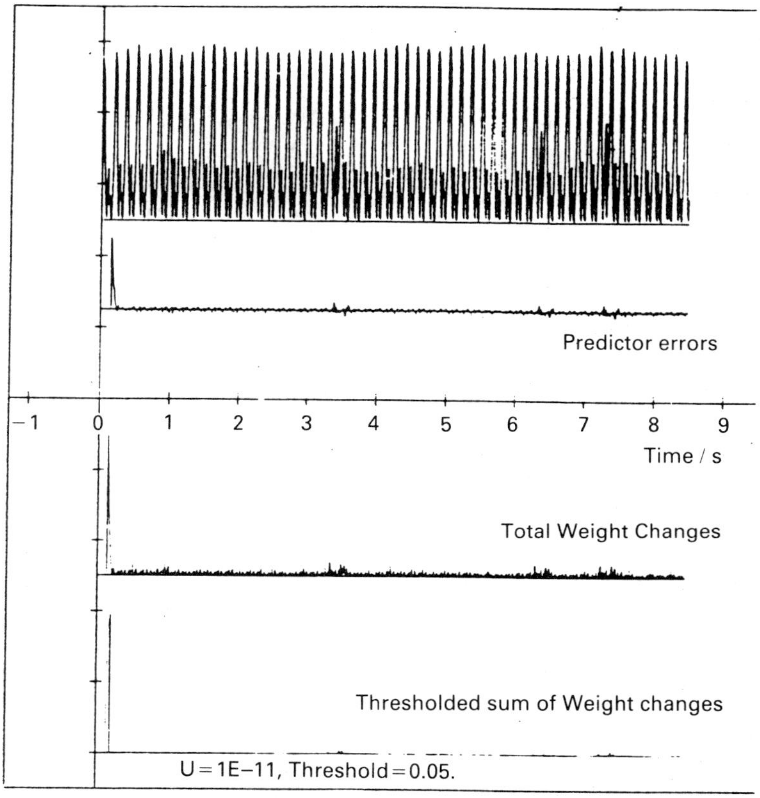

Fig. 13 — Results of adaptive linear prediction on normal stitches with thread snags.

output resulting from the noise and the natural variations in the signal smaller than that resulting from the variations due to a mis-stitch. In principle, this appears to be simple, but in practice it has proved somewhat more difficult.

It can be shown [49,52,53] that the convergence of the predictor to the minimum error state depends on the sum of a set of exponential decays whose time constants are given by

$$(T_{mse})_n = \frac{1}{4.\mu.\lambda_n}$$

Thus the rate of convergence is inversely proportional to the convergence coefficient.

The duration of the variations in the tension profile due to a mis-stitch is approximately 1 stitch period. Using a correlation coefficeint sufficiently small to avoid adapting to the mis-stitch pattern thus leads to a long convergence period compared to the duration of a stitch. Since the profile and hence the signal statistics

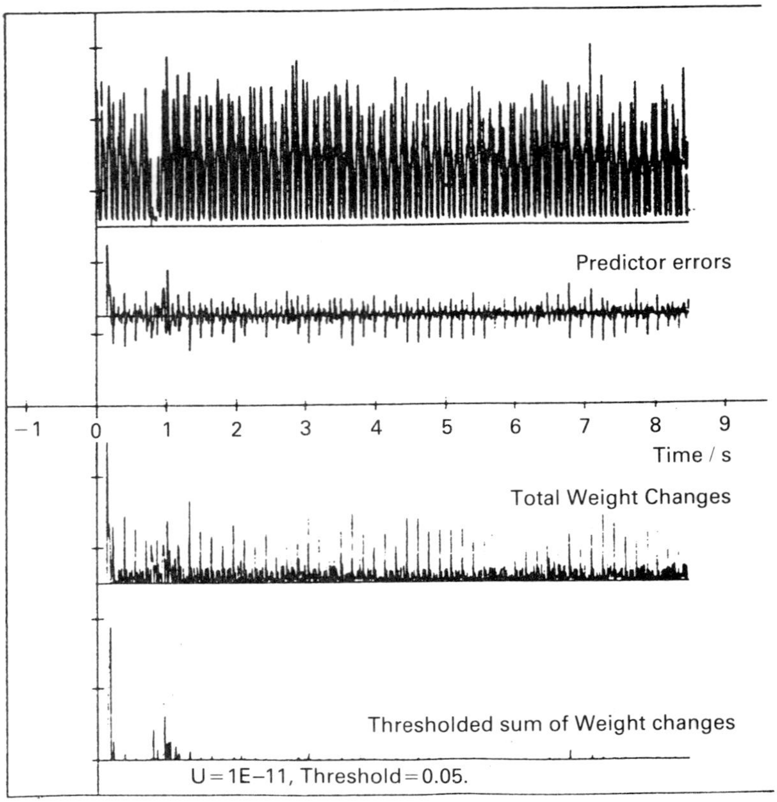

Fig. 14 — Results of adaptive linear prediction on sequence with subtle class 2 mis-stitch.

can change depending on the type of machine and its set-up, the convergence period likewise varies and in extreme cases of several dozen stitches. Since the real data only occurs in short bursts as garments are fed through the machine, a large part of each data segment would be effectively missed as the convergence took place.

Using a larger coefficient has problems too. Since the signal during a mis-stitch can drop off sharply and reduce to low-level noise, the predictor tends to track the mis-stitch, giving a smaller error during the mis-stitch. Further, during parts of a normal stitch the signal is highly variable and unpredictable and can give rise to large transitory errors. These are basically noise for the purpose of the analysis. The solution is to use adaptive techniques on the convergence coefficient, starting with a high value then reducing it as the predictor tracks in, thus giving a fast response initially and a slower one when needed later on.

The above appears to work for mis-stitch detection, but is computationally intensive. The autocorrelation discussed below may be sufficient and is much simpler. It relies on the likelihood that in normal operation mis-stitches are generally going to produce large variations of the type indicated above, rather than the more subtle variations discussed below.

Autocorrelation [48] was investigated to develop a simple and robust mis-stitch detector. The sample autocorrelation function $r(k)$, of a set of data values, $x(i)$, is defined as

$$r(k) = \frac{\sum_{i=0}^{n} (x(i) - \bar{x})(x(i+k) - \bar{x})}{\sum_{i=0}^{n} (x(i) - \bar{x})^2}$$

where k has a value between one and the number of samples, n, in the set.

The analysis involved the application of autocorrelation to single stitches. The function does not markedly differ for the normal range of variation between good stitches, but the effect of an obvious mis-stitch is sufficient to cause a major change (see Fig. 15)).

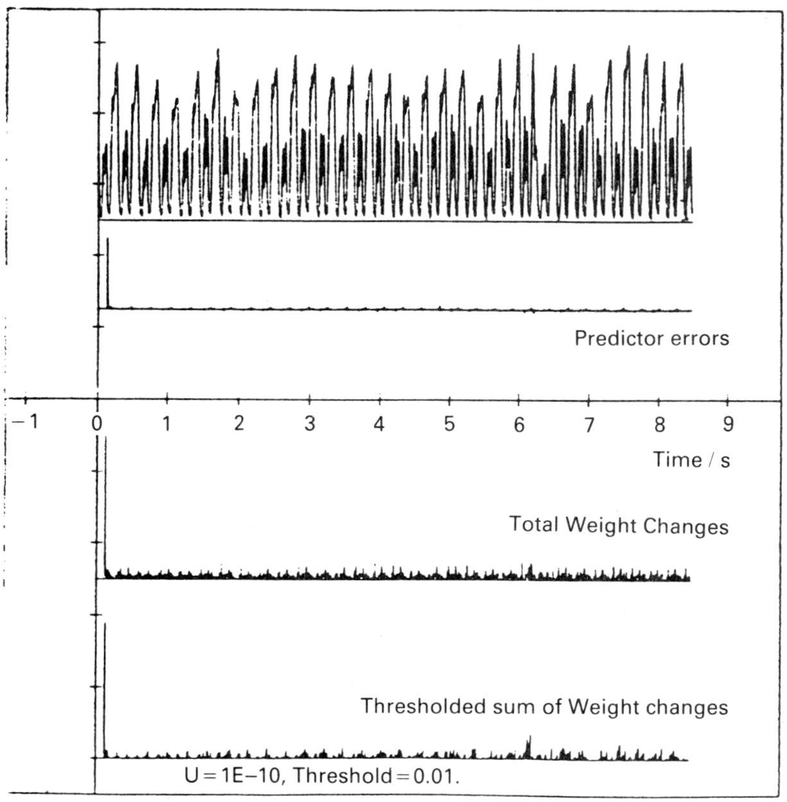

Fig. 15 — Results of adaptive linear prediction on sequence with obvious class 2 mis-stitch.

By calculating one point on the function for which a substantial difference exists between the values for a good stitch and a mis-stitch, a clear indication results. The value of the function for this delay can be calculated for each stitch and by comparing the values from a sequence of stitches to a threshold, any obvious mis-stitches will immediately show up. Subtle variations due to mis-stitches do not show up, neither do moderately large, but more transitory changes, such as those due to thread snags.

This method appears to work well, but as with the previous two, would be difficult to implement as a digital system at this time. To carry out the sampling for such a digital system would not be difficult, but to carry out the required calculations at the speed given would currently require a very expensive system. In a few years, the power of single chip processors may have increased to the point where a cheap device could handle the task. At present, given the cost limitations for a simple mis-stitch detector, such a system would be uneconomic.

The autocorrelator system, however, could be implemented in an equivalent analogue form which would be considerably cheaper.

5. CONCLUSION

A medium-sized garment factory will contain several sewing machines. The target for the simple mis-stitch detector is a system which can be attached permanently to each of these machines and it must therefore be low-cost and simple to mass-produce.

To use the results in industry, a real-time system must be developed to handle the processing for a machine running at at least 6000 rpm. For the Fourier analysis and adaptive prediction, a digital system would be needed. The autocorrelation could be implemented more cheaply using analogue techniques. Therefore, an analogue system is being developed to implement that autocorrelation.

REFERENCES

[1] T. F. P. Frank, 'A study of thread feed mechanisms in commercial high speed sewing machines', Ph.D. Thesis, Leeds University, 1979.

[2] F. C. Deery, 'The continuous measurement of thread tension during the operating cycle of sewing machines', M.Sc. Thesis, Leeds University, 1964–65.

[3] J. & P. Coats, 'Seam Pucker: why always blame the thread?', *Bobbin*, **20**, No. 3, 1978, 188–192.

[4] English Sewing Limited, 'Pocket guide to successful seaming'.

[5] C. D. Price and A. Rae, 'Variability in Sewing', British Clothing Centre Report, 21.10.1982.

[6] A. Rae, 'A Collective Responsibility. Quality Control. The Total Reality of Quality Control', *Apparel International*, **7**, No. 4, 1985, 7–8, 11.

[7] A. Rae, 'Quality Control in the Clothing Factory. Part Two', *Apparel International*, **7**, No. 5, 1985, 49–50.

[8] W. J. Blackwood and D. L. Munden, 'Factors affecting the breaking characteristics of three-thread overlock seams', *Clothing Research Journal*, **4**, No. 1, 1976, 3–34.

[9] I. Ajiki, 'Stitch tightening and compressional property of seamed fabrics for lockstitch sewing machine', *Journal of the Japan Research Association for Textile End-uses*, **29**, No. 3, 1988, 106–113.

[10] J. Gersak, 'Analysis of the stressing of sewing threads during sewing', *Tekstil*, **36**, No. 9, 1987, 481–489.
[11] P. Svoboda, 'Fault detection in large circular/knitted fabrics', *Melliand Textilberichte*, **62**, No. 2, 1981, 175–178, also **62**, No. 3, 1981, 247–251.
[12] E. Nemeth, 'Problems of the sewability of sewing threads', *Textiltechnik*, **30**, No. 7, 1980, 446–452.
[13] M. Braun, 'Influence of the sewing machine needle geometry on the sewability of knitgoods', *Melliand Textilberichte*, **58**, No. 9, 1977, 734–739.
[14] R. Nester and H. Falke, 'Changes in the strength of sewing threads during sewing on an industrial sewing machine', *Faserforschung und Textiltechnik*, **20**, No. 12, 1969, 587–593.
[15] D. H. Brain, 'Thread damage and breakage during lockstitch sewing of cotton threads', *Clothing Research Journal,* **1**, No. 3, 1973, 28–50.
[16] A. Rae, 'The characteristics of yarn tensioning devices', *Textile Institute and Industry*, 1966, 161–164.
[17] R. Parker, 'Selecting the right seam for the job', *Knitting Industry Technical Review*, **2**, No. 1, 1982, 5–8.
[18] A. Rae *et al.*, 'Report on Kirby-Lester TA7 Tension Analyser (phase 1), WIRA report ref. 1/302, February 1976.
[19] Pegasus 3-thread overlock sewing machine handbook for technical problems.
[20] Hitachi model EC1/2400 lockstitch sewing machine handbook.
[21] Ferd. Schmetz GMBH, *Schmetz Guide to Sewing Techniques*, 1975.
[22] J. Brown, Private communication. Courtaulds Clothing Brands Ltd.
[23] J. Akashi, J. Nishikawa, M. Okuno and K. Matsumoto, 'Influence of flagging on thread looping and fabric gathering', *Journal of the Textile Machinery Society of Japan*, **31**, No. 4, 1978, 165–171.
[24] F. N. Hurt and R. Parker, 'Seam geometry in knitted fabrics. I – the lockstitch seam', Hatra Research Report No. 34, March 1975.
[25] F. N. Hurt and R. Parker, 'Seam geometry in knitted fabrics. II — the two-thread chainstitch seam', Hatra Research Report No. 37, September 1975.
[26] F. N. Hurt and R. Parker, 'Seam geometry in knitted fabrics. III — the single-thread chainstitch seam', Hatra Research Report No. 41, June 1976.
[27] F. N. Hurt and D. J. Tyler, 'Seam damage in the sewing of knitted fabrics. I — Machine variables', Hatra Research Report No. 35, May 1975.
[28] F. N. Hurt and D. J. Tyler, 'Seam damage in the sewing of knitted fabrics. III — The mechanism of damage', Hatra Research Report No. 39, March 1976.
[29] F. C. Deery, N. H. Chamberlain, 'A study of thread tension variations during the working cycle in a lockstitch sewing machine', Clothing Institute Technological Report No. 15, July 1964.
[30] R. J. R. Jones and N. G. Greenberg, 'Dymanic thread tension characteristics in sewing, Parts 1 & 2, *Clothing Research Journal*, **3**, No. 2, 1975.
[31] Y. Kamata, R. Kinoshita, S. Ishikawa and K. Jujisaki, 'Disengagement of needle thread from rotating hook, effects of its timing on tightening tension industrial single-needle lockstitch sewing machine', *Journal of the Textile Machinery Society of Japan* (English Edition), **30**, No. 2, 1984, 40–49.

[32] R. J. R. Jones and D. L. Munden, 'A study of the mechanics and geometry of the 2-thread chainstitch. Part 1. An instrument for measuring static needle and looper thread tension', *Clothing Research Journal*, **8**, Part 3, 1980, 107–114.
[33] R. J. R. Jones and D. L. Munden, 'A study of the mechanisms and geometry of the 2-thread chainstitch. Part 2. The development of an apparatus for the measurement of dynamic thread tension', *Clothing Research Journal*, Part 3, 1980, 115–138.
[34] Z. Sramek, 'Faktory ovlivnujici zmeny napeti nite pri siti na sicim stroji', *Textil*, **22**, 1967, 484–488.
[35] R. Miyashita, J. Sekuguchi, M. Sasahara and H. Kandachi, 'Measuring needle thread tension and detecting skip-stitch during sewing process", *Journal of the Textile Machinery Society of Japan*, **27**, 1974, 55ff, 111–116.
[36] A. Rae and C. D. Price *et al.*, ' "Static" tests on needle thread tension system', WIRA interim report on sewing tension variations, December 1981.
[37] C. D. Price, 'Velocities and accelerations of moving parts of sewing machines', WIRA report ref. 399, May 1982.
[38] M. Disher, 'Who wants to automate? We do!', *Manufacturing Clothier*, **63**, No. 6, 1982, 31–39.
[39] 'Sewing techniques', M. Disher, *Manufacturing Clothier*, August 1987, 8–11.
[40] Patent no. US 3,731,069: 'Apparatus for detecting yarn quality information', May 1, 1973.
[41] Patent no. US 4,267, 554: 'Electronic device for monitoring a plurality of running threads on a textile machine', May 12, 1981.
[42] Patent no. CA 1,135,384: 'Control of thread processing machinery', Nov. 9, 1982.
[43] Patent no. US 4,720,702: 'Method and Apparatus for monitoring the tension of an advancing yarn'. Jan. 19, 1988.
[44] Patent no. GB 2,202,334 A: 'A method and apparatus for monitoring the tension of yarn'. Sept. 21, 1988.
[45] Patent no. US 4,377,980: 'Automatic device for sewing machines'. Mar. 29, 1983.
[46] E. O. Brigham, *The Fast Fourier Transform, and its Applications,* Prentice-Hall International, 1988.
[47] J. G. Proakis and D. G. Manolakis, 'Introduction to Digital Signal Processing', Macmillan, 1988.
[48] B. Gold and C. M. Rader, 'Digital Processing of Signals', McGraw-Hill, 1969.
[49] R. E. Kalman and N. DeClaris, eds, 'Adaptive filters' in *Aspects of Network and System Theory,* Holt, Rinehard & Winston, 1971.
[50] J. Makhoul, 'Linear prediction: a tutorial review', *Proc. IEEE,* **63,** 1975, 561–580.
[51] L. C. Wood and S. Treitel, Seismic signal processing', *Proc. IEEE,* **63,** 1975, 649–661.
[52] 'Stationary and non-stationary learning characteristics of the LMS adaptive filter', *Proc. IEEE,* **64,** 1976, 1151–1162.
[53] 'Tracking properties of adaptive signal processing algorithms', *IEEE Trans. ASSP,* **29,** 1981, 439–446.

[54] B. Widrow and J. M. McCool, 'A commparison of adaptive algorithms based on the methods of steepest descent and random search', *IEEE Trans. Ant. Prop.*, **24,** 1976, 615–637.
[55] 'A variable step (VS) adaptive filter algorithm', *IEEE Trans. ASSP*, **34,** 1986, 309–316.

QUESTIONS

SLATER: When you put your data into the system are you using a point-to-point method of detection. Are you starting and then stopping after a given time or are you continuously detecting?

CATCHPOLE: The system runs the maximum of 5 seconds or so and it then checks the results of each cycle, then waits for some time and then you get the results.

LITTLE: Thread tension I believe is affected by the thickness of the material that you are sewing as well as the type of the material, have you done any work to that effect?

CATCHPOLE: I tried looking at that and I did not get much variation at all in the samples I was using. I am not interested in the small variation.

LITTLE: I asked that question because in apparel manufacturing operations that is usually when you find most mis-stitches.

CATHPOLE: All samples were sent to me from the industry. The way that I run the experiments certainly for several thicknesses there was not much variation. I have, however, only looked at knitted fabrics not wovens.

JAYARAMAN: You've mentioned that you created mis-stitches.

CATCHPOLE: All you do is to increase the thread tension, to unbalance the stitch and by doing that you increase the probability of a mis-stitch occurring. Otherwise you have to wait for ages.

STYLIOS: On mounting of the strain gauges did you take into account the variation of the results which depends upon where you are mounting the strain gauges themselves, particularly when you are disturbing the path of the thread, and consequently not getting the representative forces as when they occur in actual sewing?

CATCHPOLE: The sensing system stands free of the sewing machine. It consists of a couple of bars with clips at the ends and the sewing thread is pulled between these and over the gauges so that the thread path above and below is not actually disturbed and hence the components of tension can be taken; you get a tight angle so you get a larger component of tension than you measured. Above and below the path of the tension is not disturbed and by keeping the metal bars polished you get as little friction as possible which is close to zero.

STYLIOS: But you are disturbing the magnitude of thread tension.

CATCHPOLE: Probably slightly.

STYLIOS: I know how difficult it is to mount a thread tension device as close as possible to the sewing needle. The other question is, when you set the sewing machine up, do you record the settings, thread tension, etc.? How do you balance the seams?

CATCHPOLE: I recorded the settings only at the beginning and then I disturbed the stitch balance to get mis-stitches.

SLATER: Did you consider the possibility of detection using either optical or capacitance techniques?

CATCHPOLE: Yes, you need a system with good resolution. Our system is cheap and it satisfies the requirement of being as close to the needle point as possible.

Introduction to Mr M. B. Gunner

The final paper in this session is going to be presented by Mr Gunner from the University of Hull. He received his B.Sc. in 1981; he worked as a research assistant in the Corah Plc project, the one that I. Gibson was talking about, and he is now working on the CIMTEX work that we are carrying out at Hull. This is a bit of a challenge to the rest of the papers in that we are not going to see any reference to sewing, considering the ratio of time spent handling to the time spent sewing is probably a little bit misrepresentative in this conference, but at least this one attempts to bridge the gap. The title of the paper is, 'Placing fabric onto moving surfaces'; it is co-authored by myself.

Placing fabric onto moving surfaces

M. B. Gunner and **P. M. Taylor**
Department of Electronic Engineering, Hull University, UK

ABSTRACT

There are many instances in *automated garment manufacture* where *static fabric panels* have to be *picked up and placed* onto a moving surface such as a *conveyor*. This paper describes the techniques used and, in particular, the effects of lowering a fabric held by its upper edge, onto a moving horizontal surface. The particular case of a *non-symmetrical piece* is analysed in detail.

1. INTRODUCTION

At present the British garment manufacturing industry is looking to automate its production in order to become more viable in the face of stiff overseas competition. It must reduce assembly costs, respond more quickly to incoming orders, reduce the amount of work in progress, and produce higher quality goods. One way of achieving this is through the automation of some or all of the production stages.

Some state-of-the-art processes are being introduced on the factory floor in an attempt to automate simple, repetitive tasks. At present cloth is laid in multiple layers before being cut into the appropriate shapes. The first job of the sewing machinist is then to remove the top piece from the stack prior to sewing. The main thrust of the work on automation has been on destacking devices, as this was an obvious starting point towards an automated system. Work has also been done on the automatic sewing of a single piece which is carried on a conveyor belt through a sewing head [1,2,3,10]. Automation is also being introduced in the manufacture of shirt collars and cuffs. These fused items are relatively inflexible and easier to handle than their constituent parts, so that automated picking off a stack and placing onto a sewing machine becomes more of a rigid component assembly problem than a flexible one. Work is also being carried out on more general problems of fabric manipulation and sewing [1,4].

Taylor [5] defines six general areas where research is required in order to produce a fully automated garment assembly line. These are:

(a) Ply separation
(b) Transportation mechanisms
(c) Orientation
(d) Pick and place
(e) Fabric manipulation
(f) Joining.

It should then be possible to create libraries of subsystems or modules in each of the above areas. Built into these modules, there is a need to relate fabric properties to the limitations of the subsystem, so that for a given garment made with a given material, it will be possible to select the required modules to produce an automated assembly line. This report deals with the problem of transportation mechanisms, by looking at the general area of placing fabric panels onto conveyors and then the more specific difficulties of placing pieces onto moving conveyors. Finally the problem of shaped panels is reviewed.

Note that all measurements are given in mm unless otherwise stated.

2. GENERAL METHODS OF PUTTING FABRIC ONTO CONVEYORS

In the factory environment subassemblies are usually passed from one work station to the next in stacks or bundles. In an automated system this tends not to be feasible as the next work station would have to remove the top assembly from the stack, then find its orientation and locate any edges required, before performing its operation. This is a difficult, time consuming, computer and sensor-dependent task. An alternative to this, which also reduces the work in progress, is to pass single subassemblies from workcell to workcell on transport mechanisms which can also be used as buffers. This passing of single subassemblies has already been implemented in some factories with the use of overhead monorail systems. Each garment on the rail is identified by an indexed position, which allows it to be tracked around the factory as it goes from operation to operation. These monorail systems can be very effective with human operatives, but are difficult to link to full automation.

Simpler systems can be implemented in the automated process by the use of conveying mechanisms which keep the subassembly in a plane. These can range from computer controlled pallets to simple conveyors. With all these approaches, the method of placing on and removing the subassembly is similar and so the general term ‘conveyors’ will be used. Conveyors can now be custom-made as there is a large selection of physical sizes, conveying surfaces and motor controllers. Once an assembly is on the conveying surface, its orientation can be disturbed by air currents or by the acceleration and deceleration of the conveyor due to the fabric’s low inertia. This can, of course, be overcome by selecting a surface with the correct frictional properties for the fabric. This may lead to a compromise, as the system will normally be used with a wide variety of fabrics.

Putting assemblies onto conveyors is dependent on whether the conveyor can be stopped for the duration of the operation. If this is so the conveying surface can be assumed to be a stationary flat plate. Many standard fabric handling mechanisms could then be used, for example, vacuum or electroadhesion. Some work on placing woven fabrics onto flat surfaces by considering their properties has already been carried out by Brown [6]. If, however, the conveyor is moving, three basic approaches can be implemented. The first is the tracking of the belt, so that a piece can be added by a flat plate gripper. (This effectively makes the conveyor surface stationary). A second method involves holding the fabric panel, parallel to the conveying surface, it can then be dropped vertically onto the conveyor. An example of this is given by Sarhadi [1], who uses a vacuum gripper to pick up fabric panels and place them onto a conveyor. A third method must be used if the handling device is holding the subassembly along one edge, allowing the opposite edge to trail free. This report will concentrate on the problems arising when using this method.

Two alternative systems can be employed depending on the relative speeds of the conveying mechanism and the manipulator holding the subassembly. If the conveyor is the faster, the subassembly can be fed onto the conveyor, in the direction of travel, by the manipulator, and held. Due to friction, the free end of the subassembly will be pulled along by the conveyor until the subassembly has been pulled out between the manipulator and the conveyor. Once the ply has unfurled, the manipulator can then release the ply. This is the 'Furl-on' Method as shown in Fig. 1. If, however, the

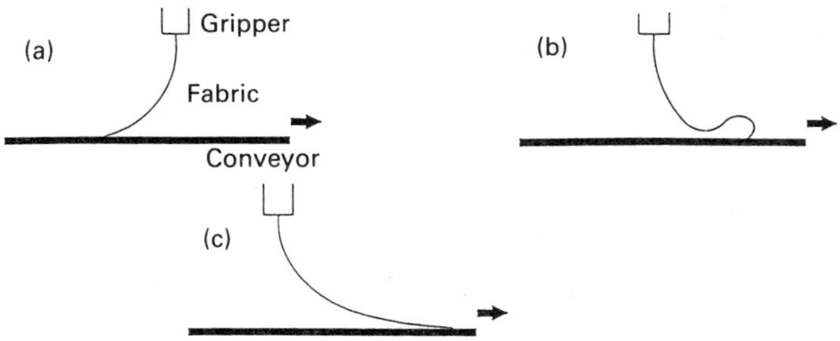

Fig. 1 — The 'furl-on' method.

conveyor is the slower, then the handling device can be used to pull the subassembly over the conveyor, in the direction of travel. Once the subassembly is completely on the conveyor, it can be released. The timing of release is crtitical to ensure that the released edge does not furl over onto the top of the rest of the assembly. This is the 'Pull-on' method. Both of these methods have been used with an automated destacking unit which has been developed at Hull University in colaboration with Corah plc, Marks and Spencer plc and Autotex plc [9]. This uses air jets to ripple the top ply on the stack of cloth. Grippers are then used to hold an edge of the fabric panel and pull it onto a conveyor belt.

3. MATERIAL PROPERTIES

It is obvious that any work on the placing of material onto conveyors will require some reference to the fabric properties, the two most important being the friction coefficient and the bending length.

3.1 Friction coefficient

Measuring the friction coefficient of fabric is more difficult than with solid surfaces as the frictional force is not independent of the geometric area in contact. Clapp [7], summarizes the published work and gives eight different methods of measuring friction. None of these methods are easy to perform, although one of the simplest is the inclined plane method, ASTM standard 3334.15 [8], which places the sample on the test surface which is then tilted until the sample starts to move. The frictional coefficient of fabric is relative to the test surface and the range of coefficient values can be large as test surface is changed. This makes reliable measurements difficult.

3.2 Bending length

The bending length of material was defined by Pierce [11] as the ratio between the fabric's resistance to bending and its weight

$$C = L\,[(\cos 0.5Y)/(8 \tan Y)]$$

where C is the bending length in mm, L is the overhang length in mm, and Y is the angular deflection in degrees.

One way of measuring this is by measuring the angle through which a specimen of cloth droops when a definite length is held over a straight edge. Unfortunately the bending of a sample is not easily defined, although Brown [6] has implemented a knowledge-based computer system which solves deflection fabric bending problems for woven material up to 10 cm long. Measurement of the bending length and relating it to the modelling of material is also difficult to perform on the factory floor.

4. THE 'FURL-ON' METHOD

4.1 First observations

In order to observe the action of placing cloth onto a conveyor, an experimental rig was produced. This consisted of a conveyor with a variable speed control. A clamp gripper (see Fig. 2) was made to hold a piece of cloth along one edge and let the opposite edge fall free on to the conveyor. The clamp gripper was used with a UMI RTX robot so that the height of the gripper above the conveyor could be easily adjusted.

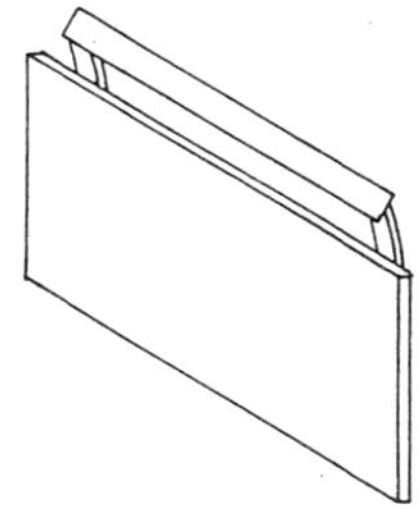

Fig. 2 — The clamp gripper.

Initial trials of the gripper involved pulling rectangular panels, 300 mm by 100 mm of cloth, which had one edge clamped using the gripper, over the edge of the conveyor surface. The conveyor surface was PVC. The robot arm moved faster than the conveyor so that once the fabric was fully on the conveyor, the robot arm was stopped and the free edge allowed to move along on the conveyor surface. This allowed the fabric panel to move as in Fig. 1. By varying the height of the held edge, it was found that lower limits could be deduced for various materials, below which unfurling would not be completed. It was also found that this height was independent of the speed of the conveyor.

The correect releasing of the fabric panel once extended was, however, dependent on the speed of the conveyor and also dependent on the height of gripper above the conveyor. By varying these settings, the released edge was liable to either:

fall onto the conveyor to make a perfect placement OR
fall onto the fabric panel, then spring back onto the conveyor OR
fall onto the fabric panel and stick.

4.2 The unfurling mechanism

Fig. 3 shows a fabric panel held at an height, *h* mm, above the conveyor. Theoretically, the curve produced by this arrangement could be calculated from detailed analysis of the fabric properties. This, however, is not easily done, so the model given here is simple and uses easily measured properties. Given that the fabric panel is of length l_t mm two other measurements can be made by holding the cloth at a set height, *h* mm above the conveyor, as if it has been unfurled (see Fig. 2(c).

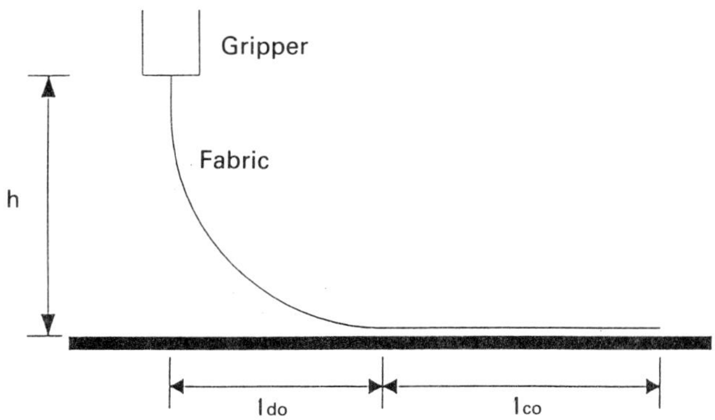

Fig. 3 — The unfurled panel.

l_{do} the distance along the conveyor, in mm, between the clamp gripper and the point where the cloth touches the conveyor.

Fig. 4 shows the cloth at its midpoint during unfurling. l_c is the length of fabric, in mm, in contact with the conveyor. l_o is the length of fabric, in mm, in contact with l_c.

Assuming that the energy to unfurl the cloth is minimal and can be neglected, then the cloth will not unfurl when the friction between l_o and l_c is greater than that between l_c and the conveyor.

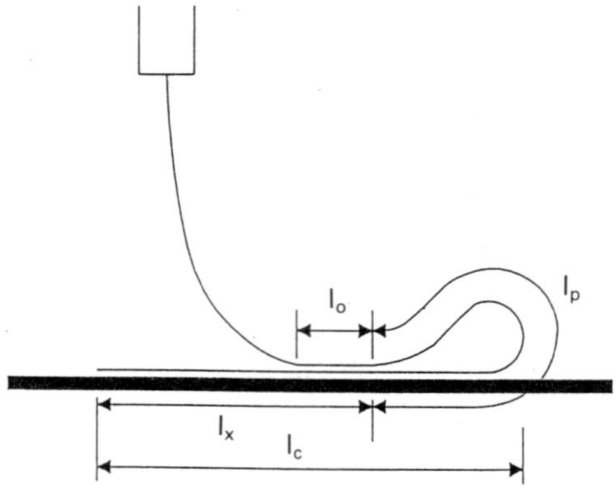

Fig. 4 — The unfurling of the panel.

Again the measurement of the frictional coefficients between the fabric and the conveyor and between the fabric and itself is difficult. A simple way to find the ratio between the coefficient of friction between the fabric and itself (Y_{ff}) and the fabric and the conveyor (Y_{fc}) is to place a panel of cloth, of length l_t, onto the conveyor surface. A second panel of the same material can then be placed onto the first so that it overlaps by a distance l_v (see Fig. 5). If the top ply is now pulled, the bottom ply will

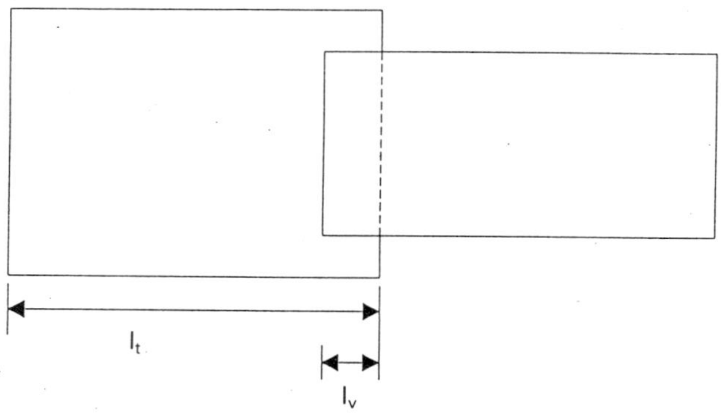

Fig. 5 — Measuring L.

move if the ratio

$$\frac{Y_{ff}}{Y_{fc}} > \frac{l_v}{l_t}$$

The distance l_v can be optimized so that when the top ply is pulled, the bottom ply just remains stationary, i.e.

$$L = \frac{Y_{ff}}{Y_{fc}} = \frac{l_v}{l_t}$$

The final measurement to be made is the length of the 'pear', l_p, which is formed as the cloth is pulled along the conveyor. A simple approximate measurement can be made folding a length of the fabric so that its two edges touch each other and placing it into a horizontal surface. The distance l_p can then be assumed to be the same length as the pear which is formed at the fold.

If the following assumptions are made, the unfurling of cloth can now be modelled.

(a) The static and dynamic frictional coefficients are the same.
(b) The force required to unfurl the cloth is minimal and can be neglected.
(c) The radius of curvature of the 'pear' is small so that $l_c = l_x + 0.5l_p$.
(d) Any furling of the edges is neglected.

An expression is required to describe when sticking will occur, in terms of the constants l_p and L.

$$L = l_o/l_c$$

$$l_c = l_x + 0.5l_p$$

$$l_o = l_{co} - l_p - l_x$$

therefore $l_x = (l_{co} - 0.5l_p(L+2))\,(L+1)^{-1}$.

But for sticking to occur

$$l_x > l_o$$

Therefore for sticking to occur

$$l_{co} > l_p(1-L)^{-1}$$

This expression allows the maximum length of fabric which can trail on the conveyor to be defined by the physical fabric constants l_p, the length of the 'pear' and L, the ratio of the coefficient of friction between the fabric and itself and the coefficient of friction between the fabric and the conveyor. Therefore, by adjusting the height of the fixed end of the fabric ply, the length of fabric trailing on the conveyor can be varied until it is less than the maximum.

The height of the gripped edge above the conveyor and the speed of the conveyor are important when considering the releasing of the fabric panel (Fig. 6). Observing the fabric panel state once released, showed that the released edge could either miss

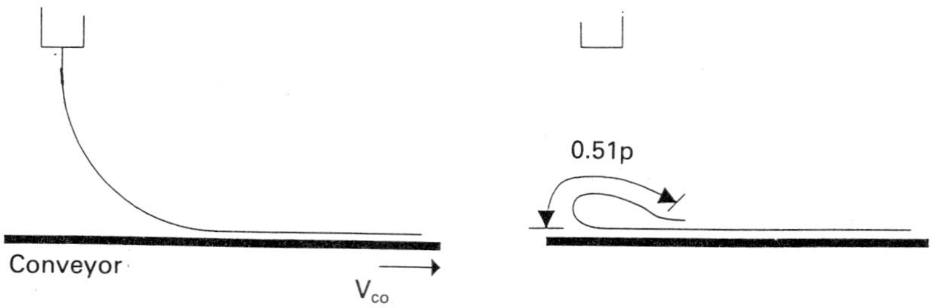

Fig. 6 — Releasing the panel.

the rest of the panel when released, or fall onto it, depending on the conveyor speed. If the edge touches the rest of the panel then, depending on the amount that touches, it will either stick or 'flip off' due to the elastic and stiffness properties of the cloth. The length of the pear l_p, can be used to define the distance that the dropped edge can fall onto the rest of the fabric panel, before it sticks.

The time, t, for the fabric to fall h mm, the distance between the clamp and the conveyor, is given by

$$t=[2h/g]^{0.5} \tag{1}$$

where g is the gravitational constant. Assuming negligible acceleration time, the distance moved by cloth along the conveyor in this time, l_c, is given by

$$l_c=V_c t \tag{2}$$

where V_c is the speed of the conveyor.

If $l_t > l_{co}+l_{do}+l_c+l_p$, then the falling edge will stick. So given l_t, h, l_{co}, l_{do}, l_p, the minimum speed of the conveyor to prevent sticking for a given height can be deduced as

$$V_c < [g/2h]^{0.5}(l_t - l_{co} - l_{do} - l_p)$$

4.3 Experimental results

Four arbitrary materials were used in these experiments:

(1) Light-weight woven cotton, used in the manufacture of shirt collars (0.010 g/cm^2).
(2) Double knit jersey, used in the manufacture of underwear (0.018 g/cm^2).
(3) Heavy-weight woven cotton, used in the manufacture of light-weight jackets (0.019 g/cm^2).

(4) Single knit jersey, used in the manufacture of ladies' underwear (0.012 g/cm^2).
(5) Brush cotton used for the manufacture of leisure wear (0.024 g/cm^2).

The three required measurements were recorded as shown in Table 1. It was found

Table 1

Fabric	l_t (mm)	L	l_p (mm)
1	300	0.15	110
2	300	0.03	70
3	300	0.10	120
4	300	0.01	30
5	300	0.09	90

that the measuring of the single knit jersey fabric and the brush cotton properties was hampered by the furling of the edges.

The unfurling mechanism was then tested using the above method and the results shown in Table 2 were obtained.

Table 2

Fabric	Theoretical l_{co} (mm)	Actual l_{co} (mm)
1	130	140
2	70	65
3	178	150
4	No repeatable results were obtained	
5	No repeatable results were obtained	

Due to the furling of the edges of the single knit jersey and brush cotton fabric, again, no repeatable results could be obtained. Stick–slip behaviour made the measurement of l_{co} difficult.

The releasing timings were also studied. The conveyor was set at different speeds and the height of the gripped edge moved until when the edge is released, it sticks on the rest of the panel. The lengths l_{co} and l_{do} were then measured (Table 3). Again, the furling of the single knit jersey and brush cotton fabric made repeatable results impossible.

4.4 Other observations

While experimenting with various lengths of rectangular fabric, an absolute minimum height between the gripper and the conveyor was observed. If the height was set to d_p the diameter of the pear, then the cloth would stick in the position shown in Fig. 1(a). This is related to the frictional coefficient between the cloth and the conveyor and the force required to unfurl the cloth. Further work must be done on this problem.

The minimum time delay, t_{min}, required between the gripper stopping with the cloth on the conveyor and the gripper releasing the cloth once it has been unfurled can be expressed as

$$t_{min}=2(Pl_{co}+l_{do})\,(V_{con})^{-1}$$

where V_{con} is the velocity of the conveyor.

It was also noted that the gripper used held the whole length of the edge of the fabric. If this was not the case, the cloth was not held at the corners so that fabric was allowed to drape on to the conveyor, then distorted results were obtained, the panel tended to 'skew' and the placement of the fabric onto the conveyor was unpredictable.

5. THE PULL-ON METHOD

5.1 First observations

An alternative method of placing fabric onto conveyors requires the gripper holding the fabric panel to move onto the conveyor faster than the velocity of the conveyor. This then drags the panel onto the conveyor, and once released, the panel will drop onto the conveyor.

Unfortunately the timing of the gripper release is critical to ensure that the fabric is placed on the conveyor without furling the released edge onto the rest of the panel.

5.2 Modelling the system

The velocity of the conveyor, V_c, must be slower than the gripper velocity, V_g, to ensure that furling does not occur. The timing of the release of the panel can be achieved in two different ways. The simplest mechanism is to wait for the gripper to extend fully, before releasing the fabric panel. Then equations (1) and (2) above still hold. The maximum conveyor speed can also be given by

$$\begin{aligned} l_c &= V_c t \\ &= V_c[2h/g]^{0.5} \end{aligned}$$

Table 3

Fabric	Conveyor speed (mm/s)	Height (mm)	l_{co} (mm)	l_{do} (mm)	Theoretical l_t (mm)	Actual l_t (mm)
1	8	130	125	65	301	300
2	8	60	170	85	302	300
3	8	150	95	85	321	300
4	8		No repeatable results were obtained			
5	8		No repeatable results were obtained			
1	125	150	110	60	326	300
2	125	70	155	80	320	300
3	125	170	80	60	338	300
4	125		No repeatable results were obtained			
5	125		No repeatable results were obtained			
1	330	170	90	60	301	300
2	330	85	160	65	283	300
3	330	190	60	60	305	300
4	330		No repeatable results were obtained			
5	330		No repeatable results were obtained			

but to stick,

$$l_c > l_{co} + l_{do} + 0.5 l_p - l_t$$

therefore, sticking will occur if

$$V_c > [g/2h]^{0.5}(l_{co} + l_{do} + 0.5 l_p - l_t)$$

The alternative method is to release the panel while the gripper is still moving. This method is obviously faster than the above, given the same gripper speed. The time taken, T_g, is defined by the velocity of the gripper and the length of the panel, l_t.

$$T_g = l_t / V_g$$

5.3 Results

The simple testing of this model involved the holding of the fabric panel by one edge and releasing it with the conveyor stationary. The results were as given in Table 4.

Table 4

Fabric	Height (mm)	$l_{co}+l_{do}$ (mm)	Theoretical l_t (mm)	Actual l_t l_t (mm)
1	100	240	295	300
2	80	260	295	300
3	120	230	290	300
4	No repeatable results were obtained			
5	No repeatable results were obtained			

These results show that the model is of the system is of the correct order of magnitude. Further work must now be carried out using a moving conveyor.

6. SHAPED PANELS

6.1 First observations

The above experimentation and modelling is based on the use of rectangular fabric panels. It is, however, rare to find such shapes in manufacturing industry, so the problem of shaped panels was then considered. Fig. 7 shows the three basic shaped used.

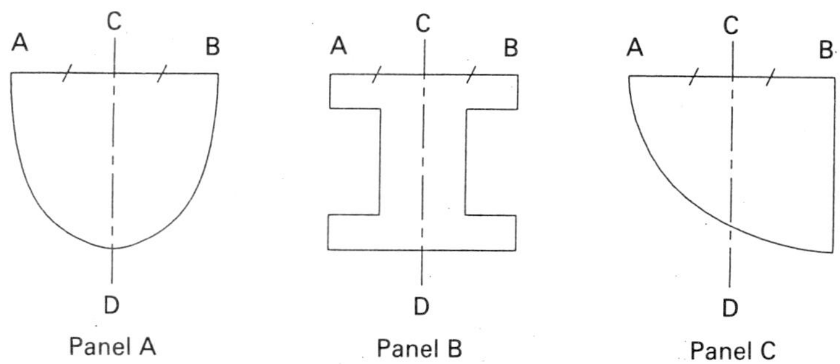

Fig. 7 — The shaped panels.

Panel A, when held along the edge AB is symmetrical about the line CD, which perpendicular to AB. Panel B is the classic I shape, while panel C is a non-symmetrical shape. Each panel was used in the experimental rig and the results noted. panel A could be placed perfectly onto the conveyor using either of the above methods. Panel B could not be placed accurately onto the conveyor. It was found that the panel deformed into the shape shown in Fig. 8. Panel C had a tendency to skew.

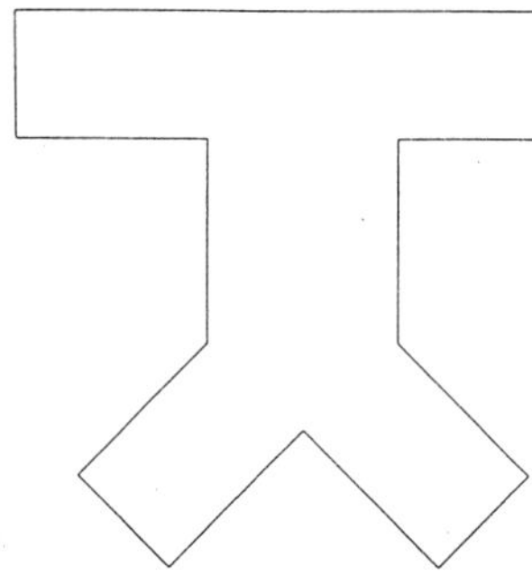

Fig. 8 — The deformed 'I'.

6.2 Modelling the shaped panels

The simplest way to model the effect of putting shaped panels onto a conveyor by clamping it by the edge is to consider the effect forces acting on one panel. Assuming that the edge of the panel is gripped perpendicular to the direction of travel of the conveyor, and the area of the fabric is left in contact with the conveyor is A. With a rectangular piece of cloth, the force on the clamp, F, is given by

$$F = Y_{fc} D A g$$

where Y_{fc} is the coefficient of friction between the conveyor and the fabric, D is the fabric density, and g is the gravitational constant.

Therefore

$$F = Y_{fc} M g$$

where M is the mass of the fabric in contact with the conveyor.

It can be seen from Fig. 9(a) that this induces 'force lines' in the cloth which are

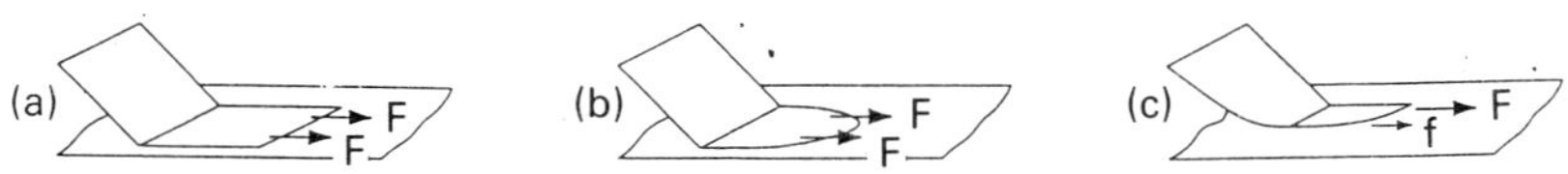

Fig. 9 — The force lines.

parallel to the direction of conveyor movement. For rectangular fabric panels, these force lines are evenly distributed across the width of the panel. For panels which are symmetrical about CD, a symmetrical pattern of force line results, shown in Fig. 9(b) and the panel can be placed onto the conveyor accurately.

Although the I-shaped piece is symmetrical and has a rectangular section on the conveyor surface, deformation of the shape occurs as there is no material along the lines of force between the edge on the conveyor and the edge being held. The result is shown in Fig. 8.

The non-symmetrical shape does not have an even distribution of force lines across the panel. The result is a 'skewing' of the panel as there is an imbalance in the forces across the panel as shown in Fig. 9(c). This skewing of the panel can be tolerated in some circumstances, for example, if a separate orientation device is to be used. If the system is automatically destacking fabric panels onto a conveyor, and directly sewing an edge, then skewing is unacceptable, so some form of counter measure must be introduced.

The simplest method is to set up a sample piece and allow it to flow along the conveyor. The gripping mechanism can then be twisted to compensate for the skewing. The Autotex destacking device uses two independent grippers to destack a panel. In order to compensate for the skewing of the panels, these can be set either to different lengths before releasing, or to different heights. In this instance, setting them to different heights would make destacking difficult, so they should be set to move out different lengths once the piece has been destacked.

An alternative method of compensating for skewing, or for different sized panels, would be to use optical sensors. These could be placed above the conveyor to look at the free edge. A feedback mechanism could then be used to alter the gripped edge before releasing. These sensors would then have to be reset for different materials. Work on this has yet to be completed.

No simple mechanism has been found to stop the 'I' shape deforming. Work must be done on how sharp the corners can be before deformation occurs.

7. CONCLUSION

This paper has outlined the problems encountered when considering placing fabric panels onto moving surfaces. It has concentrated on methods which hold the fabric panel by one edge and use the motion of either the gripper or the conveyor to transfer it onto the surface.

Two simple fabric properties have been used in these models, the first being the ratio, L, the coefficient of friction between the fabric and the conveyor surface and between the fabric and itself and the second is the length of the pear-shape, l_p, formed when folding the cloth. These are simple-to-measure constants and can be measured on the factory floor.

It was found that the model for the furl-on method produced theoretical results which were accurate to within 20% of the actual recorded results. No repeatable results were obtained for the single knit jersey fabric as the edges curled over. The model representing the release of the fabric again gave results within 20% and no results were obtained for the single knit jersey.

The pull-on method was also examined. The model for the release of the fabric panel using this method gave results within 5% of those expected, but again no repeatable results were found for the single knit jersey. The models produced for these mechanisms, although simple, have produced accurate results and could be used in a factory situation when setting up machinery. Further work is required to test the model for a wider range of fabrics. In order to make the model more accurate the differences between the dynamic and static frictional coefficients must be investigated as must the effect of the bending length and stiffness of the fabric on the pear-shape.

The problem with shaped panels has also been addressed but no simple solution has been found based on the shape of the panel. Nevertheless it was found that fine adjustment of the gripped edge before release could give accurate and repeatable placement.

ACKNOWLEDGEMENTS

The authors wish to acknowledge the following for their support: the ACME Directive of the Science and Engineering Council, CORAH plc and the DTI through the CIMTEX project.

REFERENCES

[1] Sarhadi, M., 'Practical Performance of an Automated Garment Sub-assembly Research Workstation'. Submitted to the IEEE International Conference on Systems Engineering, Pittsburg, 1990.

[2] Godrich, A., 'Semi-automatic Flat Welting Machine', Alan Godrich Sales Brochure, 1990.

[3] $(TC)^2$ National Apparel Technology Centre, $(TC)^2$ Today', Video, 1989.

[4] Taylor, P. M., Wilkinson, A. J., Gibson, I. and Gunner, M. B., 'An Integrated Garment Assembly System', Accepted for IEEE Int. Conf. Systems Engineering, Pittsburgh, U.S.A., August, 1990.

[5] Taylor, P. M., Wilkinson, A. J., Gibson, I., Gunner, M. B. and Palmer, G. S., 'Automated Fabric Handling Problems and Techniques', Accepted for IEEE Int. Conf. on Systems Engineering, Pittsburgh, U.S.A., August, 1990.

[6] Brown, P. R., Buchanan, D. R. and Clapp, T. G., 'Large Deflection Bending of Woven Fabric for Automated Material Handling', *J. Text. Inst.,* **81**, No. 1, 1990.

[7] Clapp, T. G., Timble, N. and Gupta, B. S., 'Factors Affecting Interfacial Forces Between Fabric Plies — a Survey', to be published.

[8] American Society for Testing Materials, *1987 Annual Book of ASTM Standards*, Vol. 7.01, 1987, ASTM.

[9] Kemp, D. R., Taylor, G. E. and Taylor, P. M., 'An Adaptive Sensory Gripper for Fabric Handling', Proc. 4th IASTED Symposium on Robotics and Automation, Amsterdam, 1984, 4 pp.

[10] Gershon, D. and Porat, I., 'Vision Servo Control of a Robotic Sewing System', IEEE Proceedings of International Conference on Robotics, 1988.

[11] Peirce, F. T., 'The Handling of Cloth as a Measured Quantity, *The Journal of the Textile Institute,* **20**, 1930, 377–416.

QUESTIONS

AMIRBAYAT: This work goes back 55 years, the time of Peirce, and then the work of Hearle and others. So the only properties you need is weight and thickness ratio and then bending length is the only thing in addition to frictional properties between the conveyor belt and fabric and bending length takes care of all of them. Weight and thickness ratio is there already.

GUNNER: What you are effectively doing here is you take this work and try to establish the fundamentals, which I would like fabric people to do. I do not think that at the forefront of automation we should be devoting too much time to coming up with theories of how much properties relate to each other.

AMIRBAYAT: Actually the height of your gripper is too high from the conveyor belt; that should be reduced to a minimum.

GUNNNER: You have another minimum. If you are too close on the conveyor belt the furling mechanism actually snaps at the gripper.

JAFFER: I said minimum not zero.

SLATER: Haven't you thought of not using a flat horizontal conveyor if you use a tubular one you take gravity in your side.

GUNNER: I have not actually got one of those so I haven't tried it. I think you could probably get rid of the skewing problems such as the one you are describing, but then you may have to have a different one for each material. We have to have a flexible system.

SLATER: If you develop a production line for a particular garment then you would have to use the same fabrics.

GUNNER: Assuming it is the same size, the same material. There was some work by G., Monkman with the Autotex which destacked on the first conveyor belt and then moved to the second which was lower and there were slightly sloped. The main advantage of that was to actual pull out of the material, but we did not see any extra benefits out of that system, I don't think.

DeARAUJO: Did you actually tilt the edge of the gripper to the vertical?

GUNNER: That is what I implied when I was looking at the skew panels. If you can assume that it is always picked up by the same points then you can set the machine up not by any skewing in one direction but also backwards and forwards. So if you had a complex gripper effectively on a ball joint you could have possibly got it to always be placed down properly but you will always need to pick up at the same point.

DeARAUJO: What I actually meant was tilting at an angle to the bed of the running surface.

GUNNER: The gripper is actually horizontal to the conveyor, so by tilting I do not know what benefits I would get doing that.

DeARAUJO: Have you tried changing the height during feeding?

STYLIOS: May I add to this point. It seems to me that one of the critical fabric properties here is weight. When you move your material into the conveyor belt you are really touching the edge and this is why you get the curling problem. If you were gently moving downwards reducing the height thus increasing the weight of the material, this may disturb the balance of the mechanism of friction, so that may cure the problem.

GUNNER: That is very valid and I will take care of that in further work. If anybody sees any application please let me know.

SARHADI: At high conveyor speed your result was soemthing like 20% off, I wonder if static electricity generated was considered; that is a major parameter.

GUNNER: I tend to video most of the work for 5 or 10 minutes then match it, then go back and match it again. I would not have thought that high frequency friction or electrostatics would actually build up that much so I never considered that admittedly.

Session 5: Invited discussion and general conclusions

CHAIR: Dr G. Stylios
Department of Industrial Technology, University of Bradford, UK

I would first of all like to make several observations, and then to lead an open forum for you to participate in discussions with your own observations, especially those delegates from industry, so that some feedback can be provided to us to agree in what is to be done next.

What has been revealed in these three days is as follows:

— Firstly, textile objective measurement is becoming accepted in the textile and the clothing industries, having already been accepted and used in Japan and elsewhere.
— Secondly, objective measurement is becoming established for development work on new fabrics, for tailoring of suits, for sewability problems and for routine quality assurance.
— Thirdly, there are examples of how industry can use objective measurement and how the results are implemented for its benefit.

However, there are some questions which we ought to address ourselves.

— There is a lot of work on woven suiting of pure wool and wool mixtures. Most of the presented data is associated with wool suiting material and seam pucker due to differential feed in the top sleeve area of suits. This may be due to the historical research and development involvement but if not specific to wool, evidence of results to that effect is needed.
— We ought not to neglect knitted fabrics with their inherently different to woven fabric properties and we should include woven and knitted man-made and natural fabrics in our studies.
— Textile objective measurement and automation have to be closely related to each other but the concepts here are different; we see research efforts of using this technology in computer-aided design CAD especially in the draping field, in

fabric finishing, and also in on-line measurements, especially in the optimization of sewing machine conditions. As far as fabric manipulation is concerned however and its relative problems in garment manufacture, we have not seen examples of using objective measurement in this conference.

— Consequently, it is time to re-address ourselves to the fundamental concept of flexible garment assembly and how the two areas could interact with each other to further develop this area. We saw one example in the last speaker, I also know that others like Professor Sarhadi begin to consider this technology. Fundamentally we have to consider how OMT can be used in garment automation.

I would now like to open the discussion amongst the delegates to provide us with some feedback on those points or any other points that they may like to address in this conference.

COMMENT: There are economic aspects in this area of work, as mentioned in one paper this morning. I think in a future conference, there should be some economic aspects in terms of the economic effect for the industry of some of the things that were proposed.

We had a joint conference between the Textile and the Clothing Institutes in the 1950s and it was to discuss common problems. What was then clear was that despite evident problems of garment manufacturers, fabric manufacturers replied that they could solve those problems if garment manufacturers were prepared to bear the costs. Therefore all the work done on OM of fabrics before going into an automated process it is going to increase costs and so the question is who would pay for these costs? It was mentioned that it had to go to fabric manufacturers.

STYLIOS: I do take the point, what we see is that both the textile and clothing industries use this technology to save money in the long term. I do not know the relative costing terms, but to really get into this area and to purchase one of the equipments that they are available, you can immediately justify the investment if you look at the positive results that can provide to your business. Now if you are talking about full automation, then I agree we ought to consider the economic implications seriously. However, as far as OMT is concerned, the positive results are already here before us; you reduce costs by increasing the quality of your products, you are producing added value products and consequently you would be able to become more competitive.

COMMENT: I can see the results in Japan but in the UK when you improve something others are reluctant to bear any cost of those improvements.

STYLIOS: Is it a fair criticism for the garment industry, that it is looking at a very short-term effect. To do something well you ought to look at the short as well as the long-term effects. Thus in a typical graph you would have your initial capital cost, but after that you are improving in quality and productivity. To predict the actual payback period depends upon a lot of factors, however the trend of the graph is known from Japan and other countries, why can't we see that in the UK and Europe?

DeARAUJO: The question is whether there is any other alternative.

STYLIOS: Other than closing down perhaps, how many companies have closed down in the last couple of months?

COMMENT: I serve in a rather privileged position between you and the industry; we work on consultancy within the industry as well as on certain projects, so I see both sides of the argument. Many problems that we have we can trace back to very simple mistakes in warping, spinning or in the weaving set up. It is quite surprising how often the quality is changed at these stages but the fabric still goes through to garment manufacture.

As a stop-gap measure probably OM is a way of being able to send the fabric back to the previous stage. We can predict how it can behave and then you can adjust the properties in finishing. I would like to see OM used in quality control and as a means of being able to attain certain specifications, which are nowadays getting more rigid. I do not, however, see it as the means of sending the fabric back for correction. I would like to see the material correct at the yarn stage, at the fabric stage and all the way through.

STYLIOS: Of course this is the ultimate goal for everybody, but from the garment manufacturer's point of view who has to buy fabric then here we have a very powerful tool to assess how this fabric will subsequently behave during manufacture, so I agree that if we could get it right even at the fibre point and what I refer to here is "*textile genetics*", then we would cure all the problems; however, realistically for many companies the technology is here to verify specifications and to predict the faults that they may occur and then to show how to eliminate them.

COMMENT: I agree but it is all relevant to how much you pay for the fabric. If you buy something cheap then you expect problems. OM is a way of predicting problems but if you pay more you get less problems.

STYLIOS: We have the technology that can enable us to quantify the products of this industry in a way that we were not able to do before. Does that mean that we have to turn a blind eye and say that although we all wish to use this technology we cannot because of costing restraints.

COMMENT: OMT is important in all processing sequences and if you have objective measurement as a quality control tool then you should end up with a final product which is easier to handle, therefore causing less problems and hence more economically viable to make up garments.

COMMENT: On the first day of the conference two of the speakers mentioned, almost in passing, the idea that perhaps the next step forward in the OM field could be to have on-line measurements of properties. And it seems to me that we should take this idea seriously which brings the role of OM into an entirely new perspective that we have not been addressing here. That is not to downgrade any of the papers that we heard in this conference, but they end up with what it is essentially a static measurement, bearing in mind that on-line measurements have the potential of not acting as a quality control but as process control. That then leads to the other issue I think that we have when we consider the relationship of OM with costing, with quality and desirability of the product. If on one hand we think of OM as a laboratory technique and we insist that quality of products is assessed before you go into the process and that has been the theme of the conference that's one thing, if on the other hand we use OM to actually modify the process so you can handle a wider range of fabrics than it is otherwise possible then that is a different thought. And I will be interested in this thought.

COMMENT: You can use certain equipments like the FAST for routine quality control this is one point of view, but the other point of view is the process control aspect where you look after your fabric, that is very clear in Japan especially in the presented control charts in garment manufacture.

What is very interesting is that the control charts can identify problems and can in fact be used in fabrics processed by companies supplying the Japanese and European markets and we, in Hong Kong making formal jackets, found out that these control charts were quite good. They do not include all problems that we come across but they certainly include overfeed problems and so forth. One of the interesting problems that has to be commented here is using OM in fabric construction patterns. We had a couple of fabrics where when the jacket was made, the arms in the jacket hang slightly forward on the hanger; the difference of these fabrics from all the others was that the shear rigidity and shear hysteresis were very high. When measured the pattern had to be changed to accommodate for this short of shear rigidity. These fabrics were costing the company a lot of money and the problem had to be predicted and solved.

The FAST approach is to try to make a fabric fit certain specifications shall we say, the Japanese approach is to make the process more efficient. I think the FAST approach will move this way as well.

De BOOS: I agree, but with a slight difference in emphasis; the FAST chart is designed to predict the tailoring problems, it does not necessarily demand that the fabric should perform within the control zone in the chart, we have recognized that there will be fabrics that will be placed outside the control zone for reasons of styles, for reasons of price, whatever they have to be used for. We would hope that people would use all FOM in this interpreted way not as a method of rejecting the fabrics if the cause is fabric design for instance but to recognize that problems are more difficult than others, some can be solved in tailoring and others in the fabric stage. We would note that any user of FOM should take these charts and their various forms and interpret the data in terms of the problems that they can or cannot solve.

SLATER: One factor that involves cost that has not been taken into account more than any other is the environmental cost. Whenever start looking at automation we automatically increase the energy consumption, we increase the waste, and we increase the demand of the raw material. Environmentalists and environmental matters are going to force us eventually to reduce our consumption and to reduce our quality.

STYLIOS: This is an interesting point, we unfortunately could not address it here, the idea of course was to promote OM.

COMMENT: If you get into automation and OM, you do not have seconds, power resources and so on increase.

SLATER: Natural materials as a resource will become scarce.

POSTLE: I am very pleased about the point on on-line measurement. I made this point at the beginning of my talk rather than the end because I also really think that this is the way that we must go and I was hoping that this will be picked up in this conference, not that too many people are working on it but at least the concept is here. I believe that the technology or the science at least is there on on-line measurement already and in fact the graph I have shown that tensile and shear measurements can be measured by way that can be easily incorporated on-line was taken from a paper that I wrote in 1970s when at that time I was thinking about the possibility of these things. The science is there so the principles have to be converted into something which works, of course that means more research and development and that means cooperation of various research groups and industry, this is all part of the automation process. Then of course one can carry out process control to overcome problems of return goods from customers or to construct specifications.

So to summarize, looking at the very papers presented here we can deduct that we are working at the moment in a very fragmented way. Each group is trying to solve a particular problem which is very important of course, but really the very concept means, at least up to a point, integrated manufacture, integrated in concept, that is where the research and development has to be integrated to a certain extent. And when we are talking about economic aspects of course they are important in a future conference; we should look at the economic aspects of automation, of OM, of on-line measurement and so on, because research has economic aspects too, that also is a very labour-intensive industry and it cannot be really automated because we still need human beings to carry out that process.

Now what I am leading on to is to ask people to think about how to use limited resources in a European or a world-wide basis, if you like, and consider the very limited numbers of researchers that qualify to do research and development in the textile and clothing area. There are grants available from various government and other sources but one needs to be coordinated, and to have industry support; that means that industry has to be interested in these possibilities, and to have groups to work in these areas, so that funds will come forward. Various speakers in this conference commended on the difficulty to get resources. That's something to think about. What we are really talking about is major research and development projects to be carried out by a relatively small number of groups with quite limited resources world-wide, this is not an easy task.

I would just like to express, before I finish, a couple of complimentary words about Dr Stylios; a couple of initiatives that he is taken which are part of this exercise. For example a year or so ago George initiated the *International Journal of Clothing Science and Technology*, now prior to that there was really no scientific clothing research-oriented journal in the world speaking like this. There are certainly textile journals which publish some clothing oriented papers, there have been clothing journals in the past which have disappeared and there are a couple of trade journals which occasionally publish some research and development material, but original research in the clothing area could be published but not in a clothing journal. Well that really was a major voyage in this whole area, this journal at last now provides the research and development researchers with the means of communicating. Now industrialists may not read all those papers, they don't have the time, one understands this, but that journal does need the support at least from various groups, which is necessary to promote the research and development not only in the areas that I am talking about but in all clothing areas. Now that is a very important initiative that has to be protected in the future, if we are very serious about it. The second initiative, of course, is that George is taking on this very conference which has highlighted at least some of the difficulties and pointed us to some research and development areas. At last we can go away and continue with a solid idea in what to do next. Think also about the last clothing research conference that you went to, or that you have heard about, it was probably some time ago. Therefore for the future of this whole area these initiatives are crucial and I would say that the next initiatives must somehow play a coordinated role between the various researchers and industry and all the vested interest in this area to really promote the whole idea of the clothing research and development, online, automation, etc., they are big projects and big initiatives that need some more funding. Funds are there potentially, but it is going to take a coordinated effort by academics, researchers and industry to get hold of them.

STYLIOS: Thank you Professor Postle. If I can just follow some of these comments that Professor Postle has just made.

It is clearly our wish to link with the teams of Professor Kawabata and your own team in Australia to implement and promote OM not only in the UK but also in Europe as a whole, and I think that if we can form a consortium to coordinate this particular area, something positive we would have achieved after this conference. I would be speaking to interested parties for the formation of a consortium which should consist of academics and industry.

The other thing that I have to say is that as far as the *International Journal of Clothing Science and Technology* is concerned I would also undertake to publish news and views of this discipline in a dedicated bulletin in the Journal the 'Objective Measurement Technology Bulletin for the Textile and Clothing Industries'. I will make sure that we dedicate part of it in the actual progress on OM throughout the world to reveal its present and future development in a more constructive way. Now we ought to coordinate meetings such as this for monitoring progress, exchange ideas and coordinate direction of future research and development, to that effect I would undertake the necessary measures to establish these meetings on a bi-annual basis.

LITTLE: Just to start with Professor Postle's comments and compliment you on the wonderful job that you've done for this conference. At least for me three or four very significant things occurred. The first is this whole concept of OM as the standard language of dealing between producers and users and also within the academic community and we should not be embarrassed because it's taken us a long time to recognize that there is a standard language. But now I am beginning to hear from the automation side that this may not be enough and I do not know how many others heard that but we do have a standardization and is reasonably good but there is still a little bit of division and I still feel that we need to probe that area more because some of those issues that we are not measuring should be measured and I cannot tell you more than that because I do not know. There are some of the things that some of our work is beginning to show up at the research level, I heard it from some others involved in machinery, and I don't think that we are at the end of this. And the next conference whatever you are doing I'd like to encourage you on those who are automating so that we can find out more about the fabric characteristics that are needed to really run these machines, and I am sorry I want to run these automated machines very quickly because if you make calculations and take a typical operation today that is 60% handling 40% sewing and you halve the sewing speed this is the equivalent to hiring a 70% operator and that's not good enough. Hence I would like to push that to the limit and say that if we can automate, we really want to go much further and in that area I think there are fabric characteristics that we have yet to explore, therefore I push to go to that arena with a small conference within a year and then perhaps every two years or so.

COMMENT: Basically in addition to that, the point has been made more than once in the last couple of days, but its a matter of interpretation, we do not know at the moment because we are addressing the problem from the automation aspect, we do not know if the figures that you in the OM side are producing are of use or not at the present time. Perhaps that is because we are two different disciplines. It's about time now to combine.

STYLIOS: You have hit it in the eye and that was really the theme of this conference. I am glad that we hit the bull in the eye and said that it is time now that the two disciplines became one and thank you very much for that. I think this is very important.

COMMENT: I think I would agree with most of what the speakers have said but one thing that came to me during the last day or day and a half is the amount of 're-inventing the wheel' that is taking place. And I think that it is essential that someone like yourself does form a consortium to bring together these people on a more regular basis or correspondence basis, so that we can really get some coordination of operations.

JAYARAMAN: We could actually form a network so that we can send and get information to each other world-wide; Japan, Australia, USA, etc., inexpensively.

STYLIOS: Of course we could do that under JANET in which most of us are connected to already.

COMMENT: We are also going to need better managers and I would like to quote a part of a letter from the training agency of the development of employment, in answer to a letter an applicant had written for a grant application to enable him to study for a postgraduate clothing technology course: 'This programme provides training in a wide range of subjects in the fields of information technology, new materials, advanced manufacturing technology, construction and courses in tourism and export management; unfortunately training in clothing and textile technology is not considered to be a priority area for support'. Now that is a sad state of affairs.

STYLIOS: I am afraid my position here is difficult; the only comment that I can make is that this should be directed to people who make these decisions. As far as academic institutions are concerned and especially the textile and clothing communities, I can tell you that we are trying to do our best to promote this technology, we try to do our best to make better managers, we try to do our best to get more money to do more research in order to contribute to the industry and I say that with all my heart. We are not trying to cut the mechanics out, but I am afraid we cannot make these decisions, I do not know what else we can do.

COMMENT: Yes, but despite the high level of the academic work described in this conference, it is a sad comment to say that clothing and textile technology is not considered to be a priority area.

DeARAUJO: I have an example here sitting next to me from the EC who is doing a Ph.D. in the University of Leeds without any problem in funding. There are programs like the BRITE and others.

COMMENT: I agree, but there are no mandatory grants from the local education for this area.

CLOSING OF CONFERENCE

Dr G. Stylios, Conference Organizer

I would like to thank you all most warmly, for coming to this meeting at the University of Bradford for these three days, particularly those speakers and delegates from overseas. Without your support this conference could not have taken place and without your positive, and stormy at times, participation it would not have been a success.

The mission of this conference was to promote textile objective measurement and to bridge the gaps with automation in garment manufacture. It has been fulfilled and for this I thank you all. I hope that we both academics and industrialists can see that as far as automation is concerned, it is now time to bridge the two disciplines together. If we can do that I am sure that great developments can be achieved.

It remains for me to wish you all a pleasant trip back to your countries and homes, and to wish you success in your future plans. I hope you had a pleasant and constructive time with us, and I look forward to seeing you again.

Index of contributors

Barndt, H., 107
Biglia, U., 175

Carrere, C. G., 207
Cassie, S., 47
Catchpole, J. L., 229

DeBoos, A. G., 55

Fan, J., 139
Fassima, C., 175
Fortess, F., 107
Furniss, J. C., 79, 107

Gibson, I., 191
Grosberg, P., 149
Gunner, M. B., 253

Ito, K., 81

Jayaraman, S., 163
Johns, D. J., 13
Jylhä, A., 151

Kawabata, S., 81
Kelloniemi, E., 151

Little, T. J., 207
Ly, N. G., 175

Matthews, B. A., 207
Meinander, H., 151
Munden, D. L., 47

Nitta, M., 81
Niwa, M., 81

Postle, R., 23, 27

Radhakrishnaiah, P., 163
Roczniok, A. F., 175
Ryan, L., 47

Sarhadi, M., 229
Shishoo, R. L., 119
Slater, K., 69
Stylios, G., 14, 139, 271
Sussman, N., 15

Taylor, P. M., 189, 191, 253
Tester, D. H., 55

Vass, D., 207

Wiener, M., 107
Wilkinson, A. J., 191

Subject index

Automation, 191
  algorithms, 240
  automated sewing machines, 102
  Autotex destacking device, 266
  CAD/CAM, 119
  Corah garment assembly system, 192
  intelligent environment, 142
  KESF-B1 Auto, 101
  mis-stitch detection system, 246
  overfeed sewing machine, 90
  placing fabric onto moving surfaces, 253
  problems of using conveyors for fabrics, 254, 264
  robotics, 191
  sewability system (SS), 142
  stitch quality monitoring, 229, 232
  suit manufacturing system, 88
  Swedish apparel production system (STRAPS), 119
  TEFO's production line, 123
  Y-front briefs, 191
Application of textile objective measurement, 81

Conclusions general, 271

Discussion invited, 272

Expert system, 139

Fabric, 207
  classification, 38
  comfort, 69, 71
  correlation with characteristics of, 218
  crease angle, 179
  development, 38
  dynamic and static frictional coefficients, 267
  feeding in sewing, 207
  finishing/sponging, 36, 97
  hand values, 40
  hygral expansion of wool, 124
  mechanical testing, 32, 89, 97
  objective measurement, 7
  optimization of mechanical and surface properties, 33
  performance, 55, 90
  placing onto moving surfaces, 253, 256, 264
  problems of using conveyors for, 254
  properties, 83, 86, 124, 141, 164, 179, 182, 207, 255
  pure wool crisp tropical, 39
  physical and mechanical properties, 7, 72, 120
  resin treated, 169
  shrinkage, 97
  skewing, 266
FAST, 56, 79, 107, 112
  approach, 55
  control chart, 37, 60, 112, 115
  fabric development, 38
  fabric mechanical testing, 32, 176
  interpretation of data, 57
  projects, 109
  properties, 113, 179
  reproducibility of data, 59
  tailorability, 36

Garment/Clothing/Apparel, 83, 175
  appearance, 175
  assembly system, 192
  automation, 88, 119, 191
  commercial garment manufacture, 81, 139, 191
  design, 119, 130
  drape, 134
  future of, 101
  just-in-time garment engineering, 49
  mechanics, 30
  mis-stitch detection system, 246
  optimization of sewing, 139
  prediction of garment appearance, 175
  process control charts, 83
  robotics, 191
  sewability system (SS), 139
  sewing performance, 79
  standard time determination (CMA), 123
  TEFO's production line, 123
GATT, 16

Industry based groups, 28, 40, 42, 82, 86, 97, 108, 110
Instruments, 107, 113
  fabric assurance by simple testing (FAST), 107

Kawabata evaluation system (KES), 107
mis-stitch detection system, 246
NCSU sewing dynamometer, 209
sewability system (SS), 139, 142

Kawabata evaluation system (KES), 47, 79, 81, 107, 112, 120
data chart, 40, 41, 114
fabric mechanical testing, 32, 165
KESF-B1 Auto, 101
measurements, 104, 154
projects, 108
properties, 103, 113, 163, 167
Knitted fabrics
objective measurement of, 47
physical characteristics of, 47, 50
placing onto moving surfaces, 253
seam damage, 151

Multi-fibre arrangement (MFA), 16

Predicting, 175
equation, 90
fabric conditions, 139
fabric sewability, 139
garment appearance, 175
processability of fabric, 107
seam damage, 151
seam pucker, 140
sewing machine settings and conditions, 139
tailorability, 120
Robotics, 194
applied to Y-front briefs, 191
automation, 191
Autotex destacking device, 266
Corah garment assembly system, 192
gripper, 197
placing fabric onto moving surfaces, 253
sewing constraints, 195
stitch quality monitoring, 229
UMI RTX, 191, 194, 198, 256

Sewability, 110, 139
denim, 164
fabric feeding, 211, 214
fabric sewability, 139
HATRA sew, 152
L and M sewability tester, 152
measurements, 152
mis-stitch detection system, 246
mis-stitch classes and profiles, 235
NCSU sewing dynamometer, 209
resin treated, 169
seam damage, 152, 153
seam pucker, 93, 97, 140, 197
sewability system (SS), 139
sewing performance, 163
stitch quality monitoring, 229
Sewing
automated sewing machines, 102
automatic overfeed, 90
conditions, 116
constraints, 195
difficult-to-sew fabrics, 163
dynamics, 207
dynamometer, 209
effect of speed on fabric feeding, 211
formation of 401 stitch, 234
performance, 163
predicting processability of fabrics in, 110
presser feet, 209
seam damage, 151
sewing machine settings, 139, 207
stitch formation cycles, 237
stitch quality, 229
stitch types, 231
thread tension measurement, 232
top taping of Y-front briefs, 191

Tailorability, 119
analysis and prediction, 120
FAST control chart for fabric, 36
objective analysis of, 124
process control, 111
Textile objective measurement, 23, 79
application of, 79
fabric comfort, 69
fabric mechanical properties, 119, 175
fabric objective measurement technology, 7, 27, 29
fabric quality 'fingerprint', 29
fibre properties, 69
implementation of, 40
physical characteristics, 47
present status and future potential, 27
principles of, 23
recent progress of, 81
spread of, 100

Woven fabrics, 163
difficult-to-sew, 163
fabric bipolar attributes, 28
hygral expansion of wool, 124
influence of weaving process, 164
Japanese primary hand expressions, 28
KES properties, 163
moisture absorption, 62
properties, 29, 175
pure wool crisp tropical fabrics, 39
sewing performance, 163
sportswear, 155

ELLIS HORWOOD SERIES IN APPLIED SCIENCE AND INDUSTRIAL TECHNOLOGY

Series Editor: Dr D. H. SHARP, OBE, former General Secretary, Society of Chemical Industry; formerly General Secretary, Institution of Chemical Engineers; and former Technical Director, Confederation of British Industry.

MECHANICS OF WOOL STRUCTURES
R. POSTLE, University of New South Wales, Sydney, Australia, G. A. CARNABY, Wool Research Organization of New Zealand, Lincoln, New Zealand, and S. de JONG, CSIRO, New South Wales, Australia

MICROCOMPUTERS IN THE PROCESS INDUSTRY
E. R. ROBINSON, Head of Chemical Engineering, North East London Polytechnic

BIOPROTEIN MANUFACTURE: A Critical Assessment
D. H. SHARP, OBE, former General Secretary, Society of Chemical Industry; formerly General Secretary, Institution of Chemical Engineers; and former Technical Director, Confederation of British Industry

QUALITY ASSURANCE: The Route to Efficiency and Competitiveness, Second Edition
L. STEBBING, Quality Management Consultant

QUALITY MANAGEMENT IN THE SERVICE INDUSTRY
L. STEBBING, Quality Management Consultant

INDUSTRIAL CHEMISTRY
E. STOCCHI, Milan, with additions by K. A. K. LOTT and E. L. SHORT, Brunel

REFRACTORIES TECHNOLOGY
C. STOREY, Consultant, Durham; former General Manager, Refractories, British Steel Corporation

COATINGS AND SURFACE TREATMENT FOR CORROSION AND WEAR RESISTANCE
K. N. STRAFFORD and P. K. DATTA, School of Material Engineering, Newcastle upon Tyne Polytechnic, and C. G. GOOGAN, Global Corrosion Consultants Limited, Telford

TEXTILE OBJECTIVE MEASUREMENT AND AUTOMATION IN GARMENT MANUFACTURE
G. STYLIOS, Department of Industrial Technology, University of Bradford

PREPARATIVE AND PROCESS-SCALE LIQUID CHROMATOGRAPHY
G. SUBRAMANIAN, Department of Chemical Engineering, Loughborough University of Technology

INDUSTRIAL PAINT FINISHING TECHNIQUES AND PROCESSES
G. F. TANK, Educational Services, Graco Robotics Inc., Michigan, USA

MODERN BATTERY TECHNOLOGY
Editor: C. D. S. TUCK, Alcan International Ltd, Oxon

FIRE AND EXPLOSION PROTECTION: A Systems Approach
D. TUHTAR, Institute of Fire and Explosion Protection, Yugoslavia

PERFUMERY TECHNOLOGY 2nd Edition
F. V. WELLS, Consultant Perfumer and former Editor of *Soap, Perfumery and Cosmetics,* and M. BILLOT, former Chief Perfumer to Houbigant-Cheramy, Paris, Président d'Honneur de la Société Technique des Parfumeurs de la France

THE MANUFACTURE OF SOAPS, OTHER DETERGENTS AND GLYCERINE
E. WOOLLATT, Consultant, formerly Unilever plc